Tourism Enclaves

Exclusively planned tourism destinations have increased substantially over the last decades. As a result, gated leisure communities, all-inclusive resorts, private cruise liner-owned island and other tourism enclaves are rather common features in tourism, especially in the peripheries and low- and middle-income countries. Tourism enclaves can have varied characteristics and scales of operations but typically they involve standardized 'non-local' themes or appeal in their design, activities and economies. Typically, such tourism spaces contain all or a vast majority of facilities and services needed for tourists who have limited possibilities or desires to leave the enclave. At the same time, the locals' access to these spaces is often limited or otherwise regulated. Thus, enclave tourism spaces are controlled and separated from surrounding communities.

Tourism Enclaves: Geographies of Exclusive Spaces in Tourism focuses on tourism enclaves in different theoretical and geographical contexts. The chapters of the book aim to contribute to our understanding of how these exclusive spaces are created and transformed and how they shape places and place identities.

The chapters in this book were originally published as a special issue of the journal *Tourism Geographies*.

Jarkko Saarinen is Professor of Geography at the University of Oulu, Finland, and Distinguished Visiting Professor (Sustainability Management) at the University of Johannesburg, South Africa. His research interests include tourism and development, sustainability in tourism, tourism and climate change adaptation, tourism-community relations and nature conservation studies.

Sandra Wall-Reinius is Associate Professor of Human Geography at the Mid-Sweden University, Sweden, where she is also researcher at the European Tourism Research Institute. Her research interests relate to landscape research, nature conservation, sustainability in nature-based tourism, and the tourism-local communities interface.

Tourism Enclaves

Geographies of Exclusive Spaces in Tourism

Edited by
Jarkko Saarinen and Sandra Wall-Reinius

Routledge
Taylor & Francis Group
LONDON AND NEW YORK

First published 2021
by Routledge
2 Park Square, Milton Park, Abingdon, Oxon, OX14 4RN

and by Routledge
52 Vanderbilt Avenue, New York, NY 10017

Routledge is an imprint of the Taylor & Francis Group, an informa business

British Library Cataloguing-in-Publication Data
A catalogue record for this book is available from the British Library

ISBN13: 978-0-367-71131-3 (hbk)
ISBN13: 978-1-003-14947-7 (ebk)

Typeset in Myriad Pro
by codeMantra

Publisher's Note
The publisher accepts responsibility for any inconsistencies that may have arisen during the conversion of this book from journal articles to book chapters, namely the inclusion of journal terminology.

Contents

Citation Information

The following chapters were originally published in the *Tourism Geographies*, volume 21, issue 5 (November 2019). When citing this material, please use the original page numbering for each article, as follows:

Chapter 1
Enclaves in tourism: producing and governing exclusive spaces for tourism
Jarkko Saarinen and Sandra Wall-Reinius
Tourism Geographies, volume 21, issue 5 (November 2019) pp. 739–748

Chapter 2
'Islands within Islands?' The Maldivian resort, between segregation and integration
Elena dell'Agnese
Tourism Geographies, volume 21, issue 5 (November 2019) pp. 749–765

Chapter 3
Does geography matter in all-inclusive resort tourism? Marketing approaches of Scandinavian tour operators
Sandra Wall-Reinius, Dimitri Ioannides and Kristina Zampoukos
Tourism Geographies, volume 21, issue 5 (November 2019) pp. 766–784

Chapter 4
Selling bubbles at sea: pleasurable enclosure or unwanted confinement?
Adam Weaver
Tourism Geographies, volume 21, issue 5 (November 2019) pp. 785–800

Chapter 5
Cruisers in the City of Helsinki: staging the mobility of cruise passengers
Kaisa Paananen and Paola Minoia
Tourism Geographies, volume 21, issue 5 (November 2019) pp. 801–821

Chapter 6
Airbnb as an instigator of 'tourism bubble' expansion in Utrecht's Lombok neighbourhood
Dimitri Ioannides, Michael Röslmaier and Egbert van der Zee
Tourism Geographies, volume 21, issue 5 (November 2019) pp. 822–840

Chapter 7

A place in the sun: the British housing market's (de)construction of Andalusia
Miguel García Martín, Arsenio Villar Lama and Estrella Cruz Mazo
Tourism Geographies, volume 21, issue 5 (November 2019) pp. 841–863

Chapter 8

Topologies of tourism enclaves
Richard Ek and Mekonnen Tesfahuney
Tourism Geographies, volume 21, issue 5 (November 2019) pp. 864–880

Contributors

Elena dell'Agnese is Professor in the Dipartimento di Sociologia e Ricerca Sociale at the University of Milano-Bicocca, Italy.

Richard Ek is Associate Professor in Human Geography at Lund University, Sweden. His research interest is primarily on biopolitics and spatial theory, tourism geographies and planning. Most recent publications discussed topics such as emotional labour in the sharing economy, the diffusion of micro-breweries, and overtourism.

Dimitri Ioannides is Chaired Professor of Human Geography at Mid-Sweden University, Sweden. Between 2016 and 2019 he served as the Director of the European Tourism Research Institute (ETOUR). His principal research interests relate to sustainable development and the economic geography of tourism. He is the series co-editor of the New Directions in Tourism Analysis series (Routledge).

Arsenio Villar Lama is Lecturer in the Department of Physical Geography and Regional Analysis at the University of Seville, Spain. His research interests focus on urban tourism, coastal tourism and amenity-oriented urbanization (golf), particularly in southern Spain.

Miguel García Martín is Lecturer in the Department of Human Geography at the University of Seville, Spain. His main research interests include landscape as a social construction, urban geography, heritage and its implications for contemporary culture.

Estrella Cruz Mazo is Lecturer in the Department of Physical Geography and Regional Analysis at the University of Seville, Spain. Her research interests focus on housing policies, spatial planning and residential tourism in Spain, with special attention in Andalusia.

Paola Minoia is Senior University Lecturer in Development Studies at the Department of Political and Economic Studies at the University of Helsinki, Finland. Her research focuses on international aid, environment politics, water and natural resource management, local knowledge and identities, local evaluation of environment and climate change, environmental justice, civil society dis/empowerment, environmental migration and tourism gentrification.

Kaisa Paananen is PhD candidate in the Department of Geosciences and Geography, The Helsinki Institute of Urban and Regional studies at the University of Helsinki, Finland. Her dissertation examines the mobility of tourists inside urban destinations and the effects that tourism mobility in urban environments have on other street users.

Michael Röslmaier has been working within the tourism and hospitality industry for most of his life. He received the MS degree in human geography from the Department of Tourism Studies and geography at Mid-Sweden University, Sweden, and was recently admitted into the PhD program in tourism studies in the same department.

Jarkko Saarinen is Professor of Geography in the University of Oulu, Finland, and Distinguished Visiting Professor of Sustainability Management at the University of Johannesburg, South Africa. His research interests include development and tourism, sustainability and responsibility in tourism, tourism-community relations, tourism and climate change and wilderness studies.

Mekonnen Tesfahuney is Professor in Human Geography at the University of Karlstad, Sweden. His research interest revolt around questions of precarious mobilities, power and space/place as well as human geographic and tourism theory. Recent publications dealt with the following themes: Privileged Mobilities, Pilgrimage/Post-Secular Tourism, and The Home and Tourism.

Egbert van der Zee is Assistant Professor at Utrecht University, the Netherlands, within the Department of Human Geography and Planning. His main research interests are the changing impact of tourism on cities, combining qualitative methodologies and spatial analysis of (big) data using GIS and current issues such as the sharing economy and the influence of web 2.0 on tourism.

Sandra Wall-Reinius is Assistant Professor of Human Geography and researcher at the European Tourism Research Institute at Mid-Sweden University, Sweden. Her principal research interests relate to landscape research, nature conservation, and the tourism–local communities interface.

Adam Weaver is Professor of hospitality, tourism and sport at Niagara College in Niagara-on-the-Lake, Ontario, Canada. Between 2003 and 2016, he taught at Victoria University of Wellington in New Zealand. Adam's research interests include cruise tourism, tourist behavior, tourism marketing and social theory.

Kristina Zampoukos is Senior Lecturer in Human Geography at Mid-Sweden University, Sweden. Her research interest is mainly directed towards the interrelations between mobilities and the socio-spatial formations and divisions of labour.

Enclaves in tourism: producing and governing exclusive spaces for tourism

Jarkko Saarinen and Sandra Wall-Reinius

ABSTRACT
Exclusively planned tourism destinations, such as all-inclusive resorts, gated resort communities, private cruise liner-owned islands and privatized beaches, have increased over the last few decades. Researchers have analyzed these kinds of tourism environments as enclaves, which are typically driven by external forces and actors, strongly supported by globalization and the current neoliberal market economy. Existing research shows that tourism enclaves are characterized by active border-making, power issues and material and/or symbolic separation from the surrounding socio-cultural realities, leading to weak linkages with host communities and the local economy. Tourism enclaves involve power inequalities, injustices and unsustainable practices that often have serious negative impacts on local socio-economic development. The articles of the special issue focus on tourism enclaves in different theoretical and geographical contexts and they contribute to our understanding of how these exclusive spaces are created and transformed and how they shape places and place identities. The individual research articles are contextualized and discussed with the key theoretical perspectives and empirical findings on tourism enclaves. Future research needs include analysis of linkages and flows of labor, goods, ideas and capital in different scales; policymaking, planning and regulations; environmental impacts; and locals' land and resource access in the respect of bordering, privatization and land grabbing. By focusing on these topics, the tourism industry could be guided towards more responsible and sustainable development path.

Introduction

Exclusively planned tourism destinations and resorts have increased substantially over the last few decades. These tourism spaces include different forms and kinds of tourism environments, such as all-inclusive resorts, gated resort communities, private cruise liner-owned islands, and privatized beaches, for example. In research, exclusively planned destinations and tourism environments are usually termed tourism enclaves, representing outcomes of a non-locally driven and governed neoliberal market economy (Britton, 1991; Brohman, 1996; Shaw & Shaw, 1999). However, while being externally driven they are strongly based on certain kinds of local planning and governance modes of the tourism industry and local and regional development.

In general, tourism enclaves contain all or a vast majority of the facilities, operations, and services needed for tourists and their enjoyment. At the same time, locals' access to enclaves may be controlled or even prevented, and can be limited to working purposes only. Nowadays, enclaves and enclavic processes separating tourism and tourists from local realities are relatively visible elements of tourism, especially in planning and development contexts dominated by external forces and actors (see Buzinde & Manuel-Navarrete, 2013; Saarinen, 2017; Torres & Momsen, 2005). This has led to an increasing scholarly and public interest in enclave tourism, as such forms and environments of tourism often include power inequalities and unsustainable practices that can have serious negative impacts and trade-offs for local social and economic processes (see Saarinen, Rogerson, & Hall, 2017). In global development discussions, these negative impacts and trade-offs can hinder the possibilities to use the tourism industry as a vehicle for development and a contributor to poverty alleviation, for example, in the context of the United Nations Sustainable Development Goals (SDGs) of the 2030 Agenda for Sustainable Development (United Nations, 2015).

Despite its locally problematic nature, however, enclave tourism seems to have many desirable characteristics for the global tourism industry and for organizing the value-chain optimally for revenue creation with minimal spillover, especially in foreign operational environments. Enclavic environments also seem to suit tourists, as many (and an increasing number) do enjoy being mass tourists in all-inclusive resorts (see Anderson, Juaneda, & Sastre, 2009; Butler, 1990, 1999; Wong & Kwong, 2004). In addition, recent debates in relation to uncontrolled tourism development in a form of so-called overtourism (see Milano, Novelli, & Cheer, 2019) indicate that some local residents may prefer a clearer distinction between spaces for tourists and local spaces, especially in urban residential contexts. All this makes enclave tourism a timely and highly relevant topic for research and tourism planning and governance.

This special issue takes an overview of research on tourism enclaves and aims to contribute to our understanding of how these exclusive tourism spaces are created and transformed and how they (dis)connect with local realities and shape places. In addition to theoretical understanding, the key issues relate to the implications of enclave tourism for local and regional development and their governance. After an overview of enclave tourism and related concepts and perspectives in research, the paper introduces the themes developed in this special issue by its contributing authors. Finally, future research needs are briefly discussed.

Tourism enclaves

There are different terms and definitions used for tourism enclaves and the enclavization process on a destination scale. Related terms include concepts such as the environmental bubble (Cohen, 1972) or tourist bubble (Judd, 1999) and total institutions (Dann, 2000). Regardless of the chosen term, there are two common elements in their use and definitions. First, different terms referring to enclave tourism development include a general idea that these kinds of tourism spaces are somehow separated from their surrounding environment: enclaves are based on "a well-defined perimeter [that] separates the tourism space" (Judd, 1999, p. 16), i.e. they are "isolated and separated from wider society" (Weaver, 2005, p. 168).

Secondly, their development, governance and existence are often externally driven, but take place and are processed on different scales at the same time. Shaw and Shaw (1999) indicate this non-local and multiscale nature of tourism enclaves in their definition, stating that "enclaves are operated by global capital and transnational organizations through a series of spatial networks, which unless they are strongly regulated by the local state, allow only limited economic benefits to accrue to the host communities" (Shaw & Shaw, 1999, p. 68). Enclavization can also be seen as a transformation process in which a destination evolves simultaneously toward socio-spatial homogenization and differentiation. This means that while destinations grow and transform toward similarity, resembling other large-scale resorts typical of the global tourism industry, and their original character is being eroded away in a Relphian manner (see Relph, 1976; Wall-Reinius, Ioannides, & Zampoukos, 2019 in this special issue), they are also undergoing differentiation and being separated from their surrounding socio-economic structures and characteristics (Cohen, 2006; Saarinen, 2004, see also MacCannell, 1976, pp. 44–8).

From a geopolitical perspective, the process of separating enclaves from their surrounding socio-cultural environments and regional economies can be understood as a form of human territoriality. For Sack (1986), human territoriality is a form of control that involves organizing space and creating boundaries that serve to facilitate that organization. In this respect, tourism enclaves are products of border-making between "us and them" (hosts and guests), making enclaves exclusive spaces and spaces of exclusion at the same time. On the one hand, as noted above, this means that tourism enclaves contain all or most of the facilities, services, and activities needed for visitors to enjoy their holiday in a destination (Saarinen, 2017). On the other hand, locals' access to enclavic tourism spaces is often regulated (Carlisle & Jones) with limited possibilities to benefit from tourism in terms of income and employment. As noted by Butler (1990, p. 41):

> "It is possible to almost completely avoid contact between tourists and locals [...]. Tourist enclaves can be staffed by imported labor and tourists encouraged not to venture out of the enclave, as, for example, in many casinos or Club Med-type resorts."

Thus, tourism enclaves can involve injustices and power inequalities in labor relations in a form of so-called social dumping, in which multinational companies originating from "high wage" countries search for ways to avoid legal constraints and introduce social dumping at lower levels of the production chain elsewhere or on site by using migrant workers (Terry, 2009; Ioannides & Zampoukos, 2018). Therefore, it is not surprising that many authors perceive tourism enclaves as an antithesis to sustainable development and sustainable tourism (Farmaki, Georgiou, & Christou, 2017) or inclusive growth (see Scheyvens & Biddulph, 2018). From this perspective, research has characterized enclavic spaces as international mass tourism destinations dominated and/or operated by global capital and actors providing limited benefits and participation for local communities. Thus, tourism enclaves are seen as spaces providing minimal opportunities for host-guest interaction, which widens the gap between host communities and visitors (Prayag, 2015). Enclavic tourism spaces are also regarded as having weak linkages with the surrounding regional economy, a situation that often leads to high economic leakages and low levels of local employment and salaries (see Mbaiwa, 2005).

Studies on enclave tourism

Although interest in enclave tourism has increased in the past decade, the phenomenon itself has a relatively long history in tourism development and related studies. In the 1960s, Knebel (1960) discussed attraction sites that were characterized by tourists' "own inner world" (Eigenwelt). In the 1970s, MacCannell (1973) conceptualized staged authenticity, in which host populations and/or the tourism industry constructed a front and back stage separating local realities and touristic experience needs and activity settings (see also MacCannell, 1976).

The early studies in particular focused on mapping and characterizing the tourism enclaves and the nature of connections or lack of them between the hosts and guests (Cohen, 1972, 2006; Judd, 1999). Tourism enclaves were seen as physically and mentally bounded spaces from local realities (see Lew, 2004). On the one hand, these borders – visible or mental – were porous, allowing some interaction and mobility. On the other hand, however, this interaction and mobility was often seen as one-dimensional and driven by the enclaves, the tourism industry and its needs (Edensor, 1998). This unbalanced nature of interaction was understood to be a result of the uneven power relations between the industry and local communities and actors. In the 1980s, Britton (1982), for example, critically analyzed how economically and politically powerful tourist-generating regions reorganize destination spaces by regulating and controlling tourist expenditure through limiting their mobility and excluding local producers and other local service providers from the tourism value-chain (see Shaw & Shaw, 1999).

Theoretically, the research on uneven tourism development was largely based on a dependency theory, indicating that underdevelopment is mainly a result of the geographically, socio-economically, and politically peripheral position of destination countries and regions in the world economy. For example, tourism development on islands has often been "characterized by a condition of dependency" in research (Nunkoo & Ramkissoon, 2016, p. 557). From this dependency perspective, tourism enclaves are seen as products of global capitalism (see Brohman, 1996) with vocal calls to recognize "the capitalistic nature" of tourism (Britton, 1991, p. 453). Indeed, the deepening forces of globalization, with the relatively free movement of capital and (privileged) people, have resulted in an increasing number and variety of tourism enclaves in recent decades. This critical structuralist approach paved the way to political economy and political ecology approaches in tourism development studies (see Bianchi, 2009; Mosedale, 2011; Nepal & Saarinen, 2016) that have conceptualized tourism destinations and enclaves as transnational spaces operating in a multiscalar way (see Rogerson, 1990). Torres and Momsen (2005), for example, indicated in their study that locals call Cancun in Mexico "Gringolandia," reflecting the non-local character, dominance, and governance of the destination.

Recent research has emphasized that although tourism enclaves are based on a bordering process separating the social realities of hosts and guests (Dürr, 2012), the enclaves can also be "dynamic places of negotiation, exchange, and collaboration across cultures, nationalities and classes" (Manuel-Navarrete, 2016, p. 507). From this perspective, tourism enclaves provide meeting points for the industry, local developers, and labor. These relations, however, are still often seen as being based on an uneven exchange between the tourism industry actors and communities (see Saarinen, 2017; Torres & Momsen, 2005). Timely research examples include modern slavery, in which the links between the global tourism industry's supply chains and human rights abuses in sex tourism, labor rights, and social dumping (e.g. the use of cheap migrant labor), for example, are analyzed (Cheer, 2018). Modern day slavery involves practices, processes, and institutions that undermine or restrict human freedoms, and at the same time it exploits people through illegal and/or legal ways (see Van Anker, 2004; Cheer, 2018), in which the bordering processes of tourism enclaves may play a critical role by enforcing injustices. Thus, while the borders of enclavic spaces may be conceptualized in research as permeable, negotiable, and challenged in local everyday life, the bordering process itself is often characteristically based on reinforced inequalities by the tourism enclaves and the global industry operating at different scales.

Insights from the special issue

Given the significance of borders and border-making in enclave tourism, we start this special issue with dell'Agnese's contribution. Dell'Agnese discusses the duality of borders in a Maldivian resort, where the external border regulates the movements of tourists and locals, and the inner border segregates frontline staff and other workers, usually of other origin and who are restricted in their movement. Dell'Agnese concludes that the outer/external borders are becoming more porous, whereas others remain difficult to cross.

Another significant parameter of enclave tourism is its standardized scope. The framework in Wall-Reinius, Ioannides, and Zampoukos's article is based on the idea of placelessness by Relph (1976). By investigating the marketing of all-inclusive tourism, they conclude that all-inclusive holiday resorts, through their scale, design, and function, purposefully separate themselves from their immediate surroundings. In addition, Wall-Reinius et al. further state that there seems to be a genuine lack of interest by tour operators to encourage visitors to areas and/or attractions outside the core resort structures. The authors conclude that the all-inclusive resorts pay relatively little attention to place-based attributes, representing decontextualized and interchangeable spaces. Based on their analysis the authors also suggest a typology of enclavism.

Weaver also focuses on the marketing of enclaves with a focus on the selling of seagoing tourism bubbles. In his article, Weaver highlights the co-existence of desirable aspects of cruise ships and problems that passengers encounter, such as seasickness, boredom, and discomfort. Weaver concludes his analysis based on guidebooks, magazines, and trade journals by presenting a future research agenda. He suggests that future research on enclave tourism should focus on employees and their experiences of enclosure; a deeper exploration of emotions in relation to tourist bubbles; and contemporary consumer culture and commercial structures that support and influence individual consumers. Related to the last point, comparisons between shopping on cruise ships and in malls could be fruitful research tasks.

In their study, Paananen and Minoia investigate cruise passengers leaving the controlled environment by analyzing the movements of cruise tourists in the city of Helsinki during their short visit. The authors focus on enabling and disenabling aspects of mobility among cruise passengers who choose to move independently around the city. Paananen and Minoia discuss social interaction among the travel parties, traffic culture, sense of direction, and time-related anxiety, as well as the use of wayfinding tools, such as street signs, mobile devices, and physical maps. The authors suggest that the used "go-along" research method offers a dynamic mode for observing short-term tourism, which can be useful for urban planning in tourist cities and particularly in destinations with a growing number of tourists causing disturbances and residential issues and conflicts.

Ioannides, Röslmaier, and van der Zee's article relates to the sharing economy with an example on Airbnb activities and their spatial tendencies in the Dutch city of Utrecht. According to their study, Airbnb-based economy and activities are growing in the city, and they cluster particularly in the city center and immediate surroundings of the main tourism attractions, restaurants, and other accommodation facilities (i.e. hotels). Ioannides et al. conclude that Airbnb activities take on the role as an instigator of urban tourism bubble expansion, which contributes to a gradual touristification of the urban structure, especially of the city center. According to them, future research should analyze the effects of Airbnb on housing markets, the hotel industry, social dimensions of neighborhoods, and the implications of an exodus of long-term residents as more properties transform into Airbnb units.

Based on earlier research and literature, Ek and Tesfahuney outline a conceptual article highlighting understandings of the phenomenon of tourism enclaves. They claim that tourism enclaves have mostly been understood as geographically distinct (i.e. bounded) areas. In contrast to this topographical understanding, the authors suggest topological understandings to offer more complex and nuanced conceptualizations of tourism enclaves. Based on a politico-philosophical framework and relational ontology, Ek and Tesfahuney outline ideas for future directions for research on tourism enclaves, which highlight the relational notion of space, its functionality, connections, and interactions.

The final article of this special issue deals with images selling Andalusia, Spain, to the British market. According to García Martín, Villar Lama, and Cruz Mazo, the rapid emergence of coastal resorts along the Mediterranean fringe has resulted in socio-spatial structures resembling "mini colonies," in which tourists are detached from the local physical and cultural environments. At the same time, this disconnectedness encourages a sense of belonging to their own expatriate nationality. García Martín et al. focus on the importance of elements and features in social imaginaries and discourses around Andalusia as a destination and how it is constructed in real estate websites targeting British markets. The authors conclude that the romantic projection of the image of Andalusia (i.e. its sublime, picturesque, and exotic nature) still persists, whereas other landscape qualities and settings seem to receive less visibility.

Future research needs

Enclave tourism has been criticized for its socio-economic, political, and environmental impacts and injustices. Enclaves are separated from their surrounding socio-spatial structures in different ways and on various levels, but they are not isolated from the forces and processes of globalization. Indeed, tourism enclaves are simultaneously spaces of enclosure *and* connections, and future research on enclaves should focus on the intertwined linkages and flows of workers, goods, ideas, capital, and so on, on different scales. Analysis of this enclosure - connection incongruity would create a better understanding of those driving forces, processes, and networks that lead development toward enclavic spaces on a local scale. In this respect, connections to border studies and geopolitical research would be beneficial. Fregonese and Ramadan (2015), for example, have conceptualized hotels, with their selective openness and closedness to their surroundings, as geopolitical spaces and sites.

Obviously, tourism enclaves are not totally separated from their environs. Thus, it is important to analyze in which ways the enclavic spaces interface with their surrounding communities and societies, and how benefits and costs are distributed. More research is needed on local people and their (lack of) opportunities to access certain locations, the labor market, housing, and services. Enclaves can also have impacts on wider real estate and land markets and developments. Related to this, it is important to analyze destination development policies, planning, and governance strategies and practices. In particular, local participation and stakeholder roles in policymaking and planning still remain an overlooked research topic. In addition, there may also be some positive planning and policy aspects related to tourism bubbles. From this perspective, the question is whether a clearer distinction between spaces for tourists and locals or other modes of regulations would serve a purpose in some tourism development contexts, such as conflict situations between residents and uncontrolled tourism growth. This is a relevant topic for researchers, planners and tourism developers in overtourism situations.

Enclave tourism has been associated with a range of socio-economic and cultural issues, but it may also have serious environmental consequences. These negative environmental impacts associated with enclavic tourism development can hinder the industry's capacity to act in responsible and sustainable ways, especially in peripheral areas and the Global South. Thus, there is a need for research on overconsumption, waste management, and water consumption, for example, and we suggest that future studies should highlight the consumptive aspects of enclaves and environmental pressures created by enclave tourism developments.

Tourism is a land use form and increasingly so. However, gated communities and enclavic resorts do not only occupy land; they create socio-spatial patterns of inclusions and exclusions. Thus, as manifestations of territorialization and privatization, enclaves can have serious consequences for locals' access to vital resources, such as land and clean water (see Cole & Browne, 2015). Therefore, studies on resource access and land grabbing, for example, are highly needed as they would create critical knowledge of the socio-economic, political, and spatial impacts of enclavic spaces. In this respect, the question of land grabs may have commonalities with enclosures and the enclavization process, as they are the antithesis to commons, common land and common-pool resources. In this context, future research should focus more on social and environmental justice, including resource access, equity, and the consequences of land use in tourism.

Disclosure statement

No potential conflict of interest was reported by the author(s).

References

Anderson, W., Juaneda, C., & Sastre, F. (2009). Influences of pro- all-Inclusive travel decisions. *Tourism Review*, *64*(2), 4–18. doi:10.1108/16605370910963482

Bianchi, R. (2009). The 'critical tourism' in tourism studies. A radical critique. *Tourism Geographies*, *11*(4), 484–504. doi:10.1080/14616680903262653

Britton, S. (1991). Tourism, capital, and place: Towards a CRITICAL geography of tourism. *Environment and Planning D: Society and Space*, *9*(4), 451–478. doi:10.1068/d09045

Britton, S. G. (1982). The political economy of tourism in the third world. *Annals of Tourism Research, 9*(3), 331–358. doi:10.1016/0160-7383(82)90018-4
Brohman, J. (1996). New directions in tourism for third world development. *Annals of Tourism Research, 23*(1), 48–70. doi:10.1016/0160-7383(95)00043-7
Butler, R. (1990). Alternative tourism: Pious hope or Trojan horse? *Journal of Travel Research, 28*(3), 40–45. doi:10.1177/004728759002800310
Butler, R. (1999). Sustainable tourism: A state–of–the–art review. *Tourism Geographies, 1*(1), 7–25. doi:10.1080/14616689908721291
Buzinde, C. N., & Manuel-Navarrete, D. (2013). The social production of space in Tourism enclaves: Mayan children's perceptions of tourism boundaries. *Annals of Tourism Research, 43*, 482–505. doi:10.1016/j.annals.2013.06.003
Carlisle, S., & Jones, E. (2012). The beach enclave: A landscape of power. *Tourism Management Perspectives, 1*, 9–16. doi:10.1016/j.tmp.2011.12.003
Cheer, J. M. (2018). Geographies of marginalization: Encountering modern slavery in tourism. *Tourism Geographies, 20*(4), 728–732. doi:10.1080/14616688.2018.1477171
Cohen, E. (1972). Toward a sociology of international tourism. *Social Research, 39*(1), 164–182.
Cohen, E. (2006). Pai - A backpacker enclave in transition. *Tourism Recreation Research, 31*(3), 11–27. doi:10.1080/02508281.2006.11081502
Cole, S., & Browne, M. (2015). Tourism and water inequity in Bali: A social-ecological systems analysis. *Human Ecology, 15*, 439–450. doi:10.1007/s10745-015-9739-z
Dann, G. (2000). Overseas holiday hotels for the elderly: Total bliss or total institution? In Robinson, M., Long, P., Evans, N., Sharpley, R. & J. Swarbrooke (Eds.), *Reflections on international tourism: Motivations, behaviour and tourist types* (pp. 83–94). Sunderland: The Centre for Travel & Tourism.
Dürr, E. (2012). Encounters over garbage: Tourists and lifestyle migrants in Mexico. *Tourism Geographies, 12*, 339–355. doi:10.1080/14616688.2012.633217
Edensor, T. (1998). *Tourists at the Taj. Performance and meaning at a symbolic site*. London and New York, NY: Routledge.
Farmaki, A., Georgiou, M., & Christou, P. (2017). Growth and impacts of all-inclusive holiday packages: Echoed form the industry. *Tourism Planning & Development, 14*, 483–502. doi:10.1080/21568316.2016.1272484
Fregonese, S., & Ramadan, A. (2015). Hotel geopolitics: A research agenda. *Geopolitics, 20*(4), 793–813. doi:10.1080/14650045.2015.1062755
Ioannides, D., & Zampoukos, K. (2018). Tourism's labour geographies: Bringing tourism into work and work into tourism. *Tourism Geographies, 20*(1), 1–10. doi:10.1080/14616688.2017.1409261
Judd, D. (1999). Constructing the tourist bubble. In D. Judd & S. Fainstein (Eds.). *The tourist city* (pp. 35–53). New Haven: Yale University Press.
Knebel, H.-J. (1960). *Soziologische strukturwandlungen im modernen tourismus [Sociological structural changes in modern tourism]*. Stuttgart: Enke Knoblich.
Lew, A. (2004). Editorial: Tourism enclaves in place and mind. *Tourism Geographies, 6*(1), 1. doi: 10.1080/1461668032000172283
MacCannell, D. (1973). Staged Authenticity: Arrangements of Social Space in Tourist Settings. *American Journal of Sociology, 79*(3), 589–603. doi:10.1086/225585
MacCannell, D. (1976). *The tourist. A new theory of the leisure class*. New York, NY: Schoken Books.
Manuel-Navarrete, D. (2016). Boundary-work and sustainability in tourism enclaves. *Journal of Sustainable Tourism, 24*(4), 507–526. doi:10.1080/09669582.2015.1081599
Mbaiwa, J. E. (2005). Enclave tourism and its socio-economic impacts in the Okavango Delta, Botswana. *Tourism Management, 26*(2), 157–172. doi:10.1016/j.tourman.2003.11.005
Milano, C., Novelli, M., & Cheer, J. M. (2019). Overtourism and Tourismphobia: A Journey Through Four Decades of Tourism Development, Planning and Local Concerns. *Tourism Planning & Development*, 16(4), 353–357. doi:10.1080/21568316.2019.1599604
Mosedale, J. (Ed.). (2011). *The political economy of tourism: A critical perspective*. London and New York: Routledge.

Nepal J. & Saarinen, J. (Eds.). (2016). *Political ecology and tourism*. London: Routledge.
Nunkoo, R., & Ramkissoon, H. (2016). Stakeholders' views of enclave tourism: A grounded theory approach. *Journal of Hospitality & Tourism Research, 40*, 557–588. doi:10.1177/1096348013503997
Prayag, G. (2015). Beyond Enclave Tourism: The Road to Sustainability. In N. Morpeth & H. Yan (eds.), *Planning for tourism: Towards a sustainable future* (pp. 240–258). Oxfordshire: CABI.
Relph, E. (1976). *Place and placelessness*. London: Pion.
Rogerson, C. M. (1990). Sun international: The making of a South African tourismus multinational. *GeoJournal, 22*, 345–354. doi:10.1007/BF00192834
Saarinen, J. (2004). 'Destinations in change': The transformation process of tourist destinations. *Tourist Studies, 4*(2), 161–179. doi:10.1177/1468797604054381
Saarinen, J. (2017). Enclavic tourism spaces: Territorialization and bordering in tourism destination development and planning. *Tourism Geographies, 19*(3), 425–437. doi:10.1080/14616688.2016.1258433
Saarinen, J., Rogerson, C., & Hall, C. M. (2017). Geographies of Tourism Development and Planning. *Tourism Geographies, 19*(3), 307–317. doi:10.1080/14616688.2017.1307442
Sack, R. (1986). *Human territoriality: Its theory and history*. Cambridge: Cambridge University Press.
Scheyvens, R., & Biddulph, R. (2018). Inclusive tourism development. *Tourism Geographies, 20*(4), 589–609. doi:10.1080/14616688.2017.1381985
Shaw, B. J., & Shaw, G. (1999). 'Sun, sand and sales': Enclave tourism and local entrepreneurship in Indonesia. *Current Issues in Tourism, 2*(1), 68–81. doi:10.1080/13683509908667844
Terry, W. (2009). Working on the Water: On Legal Space and Seafarer Protection in the Cruise Industry. *Economic Geography, 85*(4), 463–482. doi:10.1111/j.1944-8287.2009.01045.x
Torres, R., & Momsen, J. (2005). Gringolandia: The Construction of a New Tourist Space in Mexico. *Annals of the Association of American Geographers, 95*(2), 314–335. doi:10.1111/j.1467-8306.2005.00462.x
United Nations. (2015). *Transforming our world: The 2030 agenda for sustainable development. resolution adopted by the general assembly on 25 September 2015*. United Nations: New York.
Van Anker, C. (Ed.). (2004). *The political economy of new slavery*. New York: Palgrave.
Wall-Reinius, S., Ioannides, D., & Zampoukos, K. (2019). Does Geography Matter in All-Inclusive Resort Tourism? An investigation of the marketing approach of major Scandinavian Tour operators. *Tourism Geographies* (in press), doi:10.1080/14616688.2017.1375975
Weaver, A. (2005). Spaces of containment and revenue capture: 'Super-sized' cruise ships as mobile tourism enclaves. *Tourism Geographies, 7*(2), 165–184. doi:10.1080/14616680500072398
Wong, C., & Kwong, W. (2004). 'Outbound tourists' selection criteria for choosing all-inclusive package tours. *Tourism Management, 25*(5), 581–592. doi:10.1016/j.tourman.2003.06.002

'Islands within Islands?' The Maldivian resort, between segregation and integration

Elena dell'Agnese

ABSTRACT
The Maldives resort islands are a type of tourist enclave subject to a dual form of borderization. The islands have an external border that coincides with each island's coastline and with the limits of the private property of the resort; this border regulates the movement of tourists and locals. The islands also have an internal border that separates the outer edge of the island from the inner edge. The outer edge, which is supposed to represent the perfect landscape of the 'tropical island,' houses all of the tourist facilities, while the interior contains the structures dedicated to the metabolic activities of the resort. The frontline staff members and the tourists share the 'dreamscape' of the outer edge, whereas the maintenance workers live in the secluded space inside the island, where they are typically hidden from the sight of tourists by high walls, and their movement is usually restricted from staff designated areas to their location of work. For maintenance workers, these spaces, necessarily limited due to the small size of the coral islands, risk becoming 'islands within islands.' Recently, the Maldivian government has begun to promote projects and initiatives in support of territorial integration between the resorts and communities of neighboring islands. Thus, the outer limits of the resort islands are, today, more porous. Their internal borders, in contrast, remain very difficult to cross.

Supplemental data for this article can be accessed here.

1. Introduction

'Island spaces ... show a tendency toward the topological reproduction of the form of the island: bases and prisons on islands become islands within islands; prisons within bases become islands within islands within islands' (Mountz, 2014, p. 7). In this way, Alison Mountz tackles the mechanism of the reproduction of segregated spaces inside the spatially separated spaces of islands. Her interest lies in political segregation, but it is possible to ask whether similar mechanisms of spatial segregation may occur within tourism-dedicated islands. Minca (2009) underlined the presence of a large quantity of literature in the area of tourism studies that stresses the proximity between tourism islands (and enclavic, monofunctional tourism spaces more generally) and any 'other form of encampment and seclusion.' In the same article, Minca highlights the double-sided nature of the tourism-dedicated island, which is 'at the same time, a metaphor of a utopian re-creational space ... and a very real place, where people do real things' (Minca, 2009, pp. 88–89). The 'utopian re-creational space' is connected with 'the metaphoric deployment of "island", with the associated attributes of small physical size and warm water,' which, as Baldacchino remarked (2012, p. 55), 'is possibly the central gripping metaphor within Western discourse.' The metaphor is related to the long-standing myth of 'paradise' (Deckard, 2010) and the symbolic representations of tropical islands as 'paradises' (Baum, 1997; Edmond, 1997). The people who do real things are, of course, the workers. With regard to workers, Minca focuses his attention on the frontline service staff, and he remarks that in order to keep the dreamy quality of 'the island' landscape, individuals who work as service staff must look 'as though they are having fun,' even if they are working. Hence, their participation in leisure activities as an integral part of their work blurs the boundaries between work and leisure.

However, on a 'tourism island,' beyond the employees with customer service roles, who share the same spaces as the tourists or have, at least, a wide 'contact zone' with the tourists, there are three other groups of workers. These groups include the managers, who enjoy total freedom of movement and sometimes mix with the guests, as the captain on a cruise ship would do; there are the workers in charge of services, such as room boys and waiters in restaurants, who share tourist spaces in a subordinate position; and finally, there are the backstage workers. Both the managers and the service workers belong to the tourist imagery, whereas the backstage workers do not. Essential as backstage workers are to the functioning of the resort, the backstage workers are made invisible by restricting their movement to hidden, inner spaces in order to prevent the negative effect that the banal 'modernity' of their everyday activities (such as waste management, laundry or repair workshops) could have on the 'performative landscape' (Minca, 2009) of the 'tourism island.' For this reason, the 'tourism island' is not only a landscape where dream and reality peculiarly mix but also a place where, in order to respect the 'dreamscape' atmosphere, if certain kinds of jobs must look like fun, others must disappear. Therefore, to prevent any contact between the dream and a particular type of reality, internal separated spaces are created to conceal from the eyes of the guests both the metabolic needs of the resort and the workers engaged in fulfilling those needs. In this way, the mechanism of topological reproduction of 'islands within islands' as mentioned by Mountz (2014) is activated.

In this paper, I first present the Maldivian resort island as a planned type of tourist enclave and its establishment as a monofunctional segregated space inside the archipelagic reality of the Maldivian society. Then, I describe and analyze the resort island as an example of the 'dreamscape' of the 'tropical island,' with the help of the notion of 'iconeme,' introduced by Turri (1998) as the 'elementary unit of perception.' Subsequently, I introduce the rules regarding the use of space that are imposed on guests and different groups of workers to demonstrate how a different type of 'island' is effectively created inside the resort. Eventually, I take into account the recent attempts at breaking the economic and social segregation of this enclavic form of tourism to create territorial relations and collaborative initiatives between resort islands and the inhabited islands of the archipelago. From this perspective, I stress that these attempts at 'networking the resort' may make its external borders more porous but do not reduce the impenetrability of its internal borders, specifically of those delimitating the spaces where backstage workers work and live.

2. Materials and methods

The 'tourism island' is a metaphor that may work for different kinds of 'enclavic monofunctional tourism space[s]' (Minca, 2009). Here, the focus is on the 'resort island' of the Maldives, a region where I have had the opportunity to carry out extensive fieldwork, thanks to the presence of a research center opened by my university in one of the atolls. Maldivian tourist resorts are 'enclavic tourist spaces' open to only their guests and workers. Therefore, fieldwork is not an easy operation. The basis of this paper is research work, conducted at different resorts, which I visited several times between 2011 and 2018. I visited the backstage areas of four of these resorts. At six resorts, I undertook 'covert participant observations' and conducted visual research. In three other cases, I conducted unstructured interviews with the guest relations manager and other managers. Given the 'covert' nature of my research, I will not provide the names of the resorts I have visited; I provide the names of only those giving information through public sources (such as websites, promotional publications, newspapers, etc.). I also used data on the sizes of the islands and the structures available at all of the resorts, supplemented by maps and photos as provided by the website www.mondomaldive.it, as secondary sources.

Although each resort has a variety of features, all Maldivian resort islands have a subset of common traits that allow for generalizations. Resorts are a type of tourism enclave based on the presence of a single commercial establishment on an otherwise uninhabited island, which, because of its small size, its physical isolation, and the high level of social segregation, could be defined as 'the epitome of tourist insularity.' Because of its insularity, each Maldivian resort island is effectively surrounded by 'natural boundaries.' However, these boundaries are socially constructed, because the resort is, in every respect, a private property. According to the so-called 'one island, one tourist resort' model, which has been in place in the Maldives since the 1970s, only one resort can be built per island, and the area contained by each resort must be the same as that of the entire island. Thus, the resort area corresponds to that of a portion of the space (the island) rented to private entrepreneurs by the Maldivian government for tourism development. In addition to being an enclave for physical and legal reasons (an island managed as a private property that can be accessed by only paying guests and laborers), the resort island has for decades been the only Maldivian area open to international tourism. In fact, enclavic tourism was imposed by the Maldivian government in the 1980s to keep tourism 'out' of the social body of the national community, and this same approach was implemented as the only possible path for tourism development until the first decade of the new millennium. At that time, the development of a new attitude in tourism regulation opened the doors to other forms of tourism and new ways of integration between resorts and inhabited islands. Set up within a previously unused area (a 'virgin' island, as suggested by tourism communication), the Maldivian resort island in its scenery juxtaposes the 'paradisiac dreamscape' of the 'tropical island' (Baldacchino, 2012; Connell, 2003; Minca, 2009), with the 'architecture of pleasure' of the 'luxury retreat' (Ayala, 1991; King, 1997). In addition, due to the absence of a permanent Maldivian population, each resort island works as a 'space apart,' where the norms imposed on the rest of the archipelago by the Islamic religion are ignored, both by international tourists and by the workers of non-Maldivian origins. Even the time zone is different: to offer increased hours of daylight for the guests and to shift the sunset toward a more 'European' time (7 pm, instead of the equatorial 6 pm), the clock is set one hour ahead of the national time.

3. Resort Islands as tourist enclaves

There is a wealth of literature on tourist enclaves (Aili, Jiaming, & Min, 2007), and a detailed examination of this body of work is outside the scope of this paper. Here, I compare the main points of the debate with the characteristics of the Maldives resort island to identify what aspects this type of resort island has in common with what is described elsewhere in the literature, and where it differs. The Maldivian resort islands belong to the category of resort enclaves (Cohen & Neal, 2012) and usually may be described as 'top-of-the-range properties' (King, 2001, p. 175). Similar to many other resort enclaves, the Maldivian resort islands are 'fenced off and sterilized' spaces (Krippendorf, 1987, pp. 70–71), free of any possible contamination by the 'real society' outside of them. At the same time, the Maldivian resort islands offer their visitors a 'carefully constructed exotic landscape of otherness' (Carlisle & Jones, 2012, p. 10). Moreover, similar to all other tourist enclaves, Maldivian resort islands have a limited economic effect on the local community because they usually exclude local service providers and limit the mobility of tourists, preventing inhabited local islands from receiving any benefit from tourist economies. Indeed, 'each resort has its own island and supporting infrastructure and imports the bulk of its labour as well as consumption goods from abroad' (Ministry of Tourism Arts and Culture, 2011, p. 11). Thus, the Maldivian resort islands fit into the category of those 'gated communities of tourism,' which 'with their all-inclusive products can turn out to be all-exclusive spaces for local communities in development' (Saarinen, 2016).

In addition to the guests, the only people admitted to the Maldivian tourist resorts are usually the workers. In general, as the literature suggests (MacCannell, 1992), expatriates of (mainly) European origin often take positions of responsibility and top-level technical positions, such as chefs, accountants, and diving instructors, while the locals occupy the low-paid jobs. Unlike elsewhere, however, in the Maldives, the locals are not 'squeezed out of every role except the menial positions that they have always occupied' (MacCannell, 1992, p. 174). Instead, the best qualified locals occupy positions of responsibility or share the better-paid jobs with expatriates of mainly European origins, while the lower-level positions are mostly taken by expatriates of South Asian origin. Specifically, in the case of low-skilled occupations, typically, the Maldivians prefer roles such as room boy or waiter, which places them in direct contact with the guests (and their tips) (Shakeela & Cooper, 2009), while foreign workers from South Asian countries carry out the most demanding jobs, such as cleaners, gardeners, and laborers, typically without any contact with the public.

Finally, Britton points out that the development of tourist enclaves in peripheral economies is generally dominated by central economies, as the control of accommodations, means of transport, airlines, banks, and shopping malls by large foreign companies leads to the formation of enclaves (Britton, 1982). From this perspective, tourism enclaves and resorts are often accepted as the 'products of global capitalism and a non-locally-driven neoliberal market economy' (Saarinen, 2016); if related to certain kinds of tourism planning in destination societies, such enclaves are the results of a neoliberal governance 'which emphasizes the role of the markets rather than the state in guiding and controlling the development discourses and practices' (Saarinen, 2016). In the past, some low-income countries cultivated the idea of maximizing 'income from the industry by encouraging mass tourism with a minimum of local contact' (Lea, 1988, p. 24) and promoted the creation of tourism enclaves. This is what happened in the Maldives, where the segregation of tourist spaces was imposed in the 1980s by the local administration and remained the only option for tourist development until the first decade of the new millennium.

4. The framework: the Maldives as a tourism destination

Located in the Indian Ocean, near the Equator, the Maldives is an archipelagic state of 1,192 coral islands which extends within an EEZ of approximately 859,000 km^2. On average, the Maldives Islands are very small and have surface areas measuring less than 5 km^2; only a few are larger than this size and exceed 2 km in length. All the islands also have a very limited altitude of less than 2 m above sea level. These physical characteristics, combined with the tropical environment, give the islands their 'Robinson Crusoe factor' (Butler, 1993), i.e. their most famous tourist attraction (Scheyvens & Momsen, 2008; Shakeela & Weaver, 2017). Indeed, 'the charm of small, fragile and delicate coral islands is the main feature that differentiates and enhances the exclusive image of the Maldives on the market,' as the *Maldives Third Tourism Master Plan 2007–2011* (Ministry of Tourism and Civil Aviation, 2007, p. 23) underlines.

The capital island, Malé, is only 5.8 km^2 in size, although the island houses approximately 38% of the 352,795 residents of the country (National Bureau of Statistics, 2017).

The remainder of the population is scattered throughout the other 187 inhabited islands of the archipelago. The overcrowding of the island capital is countered by the limited demographic size of the so-called local islands, many of which have fewer than 500 inhabitants. The 1,004 islands that do not house a resident population are, in national statistics and censuses, defined as 'non-administrative islands.' Insular monofunctionality is a peculiar tradition of the archipelago (Rufin-Soler, 2005). Therefore, specific functions are dispersed among the islands. There are airport islands (such as Hulhulé) and prison islands (such as Dhoonidhoo in the Kaafu atoll); there is a big artificial garbage island near Malé (Tilafushi), and there are approximately 120 islands exclusively leased for agricultural or industrial purposes. Tourist resorts occupy the remaining 126 islands (National Bureau of Statistics, 2017).

From a cultural point of view, the Maldives has traditionally been a fairly unified country, with only one spoken language, Dhivehi, and only one religion, Islam. Until the 1970s, even if all Maldivians accepted themselves as Muslims, women did not wear the veil. In the following decades, the nation has experienced some important changes. First, a more rigorous interpretation of religious practices has been progressively introduced (in 1997, Islam was designated the only state religion, and in 2008, a revision of the Constitution, Article 9, section D, imposed the adherence to the state religion by every Maldivian citizen). This means, among other things, that dogs are prohibited and that products not permitted by Islam (such as alcohol, pork, pornography) or symbols of other religions may not be owned or imported. The second factor of transformation is the development of international tourism, which, in the same time, has become increasingly prosperous and has replaced fishing as the country's main economic sector. As a result, if 'in the early 1980s, [the Maldives] was one of the world's 20 poorest countries… today, it is a middle-income country' and can be defined a 'development success story' (The World Bank, 2013). However, economic growth has not been homogeneous within the country, which now seems to suffer from growing inequalities, low levels of economic competitiveness, and high youth unemployment (UNDAF, 2015). Unpacking data at the regional level also reveals inequitable opportunities across the Maldives, between Malé and the atolls, mostly in terms of education and income (Sobir, Shiuna, Ibrahim, & Shafeeq, 2014, p. 27). A third factor of change is linked to the growing presence of foreign labor, which is related to the development of the construction sector and tourism. According to the 2014 census, there are '60,030 resident foreign nationals' in the country (National Bureau of Statistics, 2014). Approximately 2,000 of these foreign nationals are of European origin, but most are from South Asia. Specifically, there are approximately 13,000 people from India and approximately 7,000 from Sri Lanka. Migrants from Bangladesh represent the vast majority (37,000, of whom 36,517 are males). They are mostly unskilled workers, and in addition to the construction industry, these foreign nationals work primarily in tourism.

Today, tourism provides, both directly and indirectly, approximately 78% of total GDP (World Travel & Tourism Council, 2015). Visitors to the Maldives are mostly attracted by the 'tropical island' cliché. According to the *Maldives Visitors Survey* (Ministry of Tourism, 2017a), 'trend analysis from October 2013 to February 2017 surveys show that the beauty of white crystalline beaches and rich coral reefs of Maldives have remained constant as the top two motivations to visit the Maldives.' In the 2017 Survey, nearly 19% of international visitors said that they had visited the Maldives for its beaches, 16% of visitors for its underwater beauty, 10% for its small islands, and 6% for the privacy guaranteed by the exclusivity of the resorts. For most of these visitors, the image of the Maldives as a luxury destination was also attractive. According to the same source, the typical total holiday expenditure for each individual tourist is often quite high: 32% of respondents spent between $5001 and $10,000, 12% between $10,001 and $20,000, and 3% more than $20,001.

The total number of arrivals (1,286,135 in 2016) (National Bureau of Statistics, 2017) is approximately four times higher than the volume of the local population, and in the five years between 2012 and 2016, this value has shown a trend of stable and steady growth. In 2016, there were 37,482 beds available through the tourism industry, 72% of which were concentrated in the 126 resort islands, while the other 28% were scattered among hotels, safari vessels, and guesthouses. Resorts, all rated from 3 to 6 stars, are typically branded as luxury destinations and attract 73% of the country's international visitors (Ministry of Tourism, 2017b).

Resort islands are state-owned and leased to private parties: 44.4% are owned by local companies, while the rest are held by foreign companies (37.5%) or joint ventures (18.1%) (Ministry of Tourism, 2016). The share of tourism revenue that does not go into local hands is quite large (it can be up to 60–80%, as reported by Sobir et al., 2014, p. 56). In addition to foreign ownership and management, the two main reasons for this fact are the import of consumer goods and the employment of a significant proportion of the expatriate workforce in resorts (despite government regulations stipulating that expatriates should not exceed half of the total number of employees).

The massive presence of foreign employment in the tourism sector is due to several factors. These include the fact that Maldivians are often not sufficiently qualified for the better-paid positions, such as chefs, diving instructors, and accountants, and their lack of interest in the most humble jobs, such as cleaning staff, gardeners, and kitchen assistants. Moreover, the 'Maldives is located in a region where there is a large docile labour force; hence importation of labour is relatively cheap and easy' (Ministry of Tourism, 1996, p. 48). Therefore, 'it is possible to find expatriate labour by offering much less (in the range of US$150.0 to US$200.0 per month for unskilled labour) compared to what is usually demanded by locals to perform the same tasks' (Mohamed, 2013, p. 14). As reported by the *Maldives Tourism Master Plan 1996 – 2005* (Ministry of Tourism, 1996):

> the demands of foreign workers are lower … because of the lower living cost and limited alternative economic opportunities at their home country for them. Therefore, wages in resorts are the price expatriate labour demands, which is not attractive enough for a Maldivian.

In addition to requiring significantly lower salaries than the locals (Mohamed, 2013), expatriate staff members will accept meals and accommodations that do not meet the expectations of the local, Maldivian staff (Ministry of Tourism Arts and Culture, 2011). For this reason, expatriate labor is in great demand.

Unlike resorts, guesthouses are mainly owned and run by locals and aim to attract mid- to low-end market segments. Due to the costs and difficulties of travelling between atolls, guesthouses are mostly located near Malé or on other islands that are easily accessible by public transport (Zubair & Bouchon, 2014). Because guesthouses are situated on inhabited islands, they cannot provide their guests with the 'exception' status guaranteed by resorts with regard to religious standards regarding alcohol and clothing. However, guesthouses often manage to reserve a small part of the coast of their island as a 'bikini beach' for tourists.

5. The 'one Island, one resort' strategy

The development of tourism in the Maldives began in 1972, when the first group of visitors arrived in the archipelago, following the arrival of an Italian-English traveler named George Corbin, who wanted to launch tourism on the islands. In October of the same year, Corbin opened, in joint venture with some Maldivian friends, *Kurumba*, the first resort. *Kurumba* was built on Vihamanaafushi, an uninhabited island near Malé that had a coconut plantation. In December of that year, another group of entrepreneurs opened the second resort. That resort was called *Bandos*, and once again, the resort was the single commercial activity on an otherwise uninhabited island. By 1979, there were 25 operating resorts, all organized following the original prototype of a single tourism resort on a 'virgin' island (the so-called 'one island, one resort' model). As noticed by a group of observers from the World Bank (Lateef, Loos, Santen, Doggart, & Tanaka, 1980, p. 46): 'each resort occupies its own island and forms a self-contained tourist community. There are no Maldivian settlements on tourist islands other than to accommodate staff.' The separation between tourists and locals was later imposed by law; in the early 1980s, the Maldivian government forbade tourism on local islands, with the triple aim of protecting the market for the resorts, ensuring the quality of the tourism offer, and preserving the cultural specificity of the inhabited islands.

These early years were a period of laissez faire, but at the end of the 1970s, tourism development was placed under national control. In 1978, a new Department for Tourism and Foreign Investment (DTFI) was created (renamed Department of Tourism in 1982 and turned into Ministry of Tourism in 1988); in 1979, the main law regarding tourism (Act no. 15/79, Law on Tourism in the Maldives) was passed, and the basic regulations for the resorts concerning registration, operation, and tax policies were outlined. However, at that time, resorts were not the only tourist accommodations in the country, because a handful of guesthouses were now active in the islands (Niyaz, 2017). The situation changed markedly when, 'according to the government authorities, "to find a solution for this socio-cultural problem and to preserve the indigenous values of locals living in inhabited islands," the government withdrew the permits of guesthouses (May 1, 1984)' (Niyaz, 2017, p. 98). The ban on guesthouses de facto ghettoized international tourists, preventing them from sleeping anywhere except in resorts; at the same time, islands developed for tourism remained off limits to local people. The ensuing tourism segregation was the product of different processes and demands; from one perspective, as observed by Niyaz (2017), many entrepreneurs feared that if tourists started to go en masse to the guesthouses, the market for their resorts would be reduced. From the other perspective, segregation was a response to the independent behavior of tourists' and their standard of dress (which was considered 'disrespectful of local customs and religion,' according to Scheyvens, 2011, p. 152). Moreover, another reason offered for controlling the development of the sector was connected with the desire to avoid any deterioration in the quality of services by, for example, opening doors to an excessive number of visitors (Firaag, 1997, p. 187).

This tourism development strategy has been regarded for years as a highly successful model by both independent scholars (Kundur & Murthy, 2013; Saeed & Annandale, 1999) and international organizations, such as WTO (see Scheyvens, 2011; Shakeela, Ruhanen, & Breakey, 2011). The strategy of spatial segregation of tourism was praised 'for the positive impact on tourism profits and the quality of tourist resorts,' although, as Sathiendrakumar and Tisdell admitted as early as (1989, p. 263), it was already highly apparent that 'almost all tourism expenditure ends up in the accounts of tourist resorts.' The local model of the tourism enclave was also praised for its ability to avoid cultural contamination.

> Because of the "resort island" tourism development policy, the Maldives was able to reduce the social cost of tourism… it is claimed that the demonstration effects of tourist consumption patterns create new tastes and stimulate the local demand for foreign consumer products. However, it could be argued that this effect is minimal in the case of the Maldives as 70% of the population lives outside Malé atoll. The policy of isolation of foreign tourists in the Maldives has reduced the risk of such a chain of events

(Sathiendrakumar & Tisdell, 1989, p. 259). Therefore, the spatial segregation of tourism remained the main strategy for planning in the sector, as dictated by the new Maldives Tourism Act (Act No 2/99) and subsequent National Development Plans (NDPs), the *First Tourism Master Plan (1983–1992)*, the *Second Tourism Master Plan (1996–2005)*, and the National Environmental Action Plans (NEAPs) until the 2000s.

Over the years, the Maldives' tourism development policy issued a series of regulations on the environmental and functional characteristics of resorts. Specifically, the *First Tourism Master Plan* (1983–1992) 'recommended setting a 20% built area percentage, maximum height of buildings concealed within the vegetation and 5 m linear beach for each room as minimum standard' (Ministry of Tourism Arts and Culture, 2013, p. 221). In addition to controlling the aesthetics of the resorts and their built environment, the Maldivian government imposes strict regulations on their environmental qualities and carrying capacity limits, island vegetation, solid waste, wastewater production and recycling practices, water consumption, and energy consumption. Furthermore, the opening of any new resort is subject to a mandatory environmental impact assessment (EIA). Hence, the Maldivian agenda toward tourism development has been acclaimed as 'a very suitable form of tourism, appropriate for the small island environment' (Kundur & Murthy, 2013, p. 50). Even the rigid enclavization of tourism, which ensued from the 'one island, one resort' pattern and from the guesthouse ban of 1984, has been celebrated in environmental terms, because 'the net environmental effect of resorts and their use reflects a dynamic balance between harmful and beneficial impacts' (Price & Firaq, 1996, p. 93).

6. The resort Island as a 'dreamscape'

The Maldivian resort island shares different modalities of 'post-modern spatiality' with other types of tourism enclaves (Minca, 2000, p. 398). First, these commonalities are the result of a process of 'de-differentiation and specialization of the resort product,' aimed at offering visitors, through recurring themes, the idea of entering a space of pleasure (Shaw & Agarwal, 2012, p. 245). This process of de-differentiation and 'resortization' of the spaces open to guests is explained by the desire to, on the one hand, provide quality services and, on the other, to adapt the context to the expectations of tourists. Thus, 'the carefully constructed exotic landscape of otherness (hotels, internet-cafes, exotic restaurants, souvenir shops and craft stalls) promotes a sense of ease and security and a unique experience of relaxation' (Carlisle & Jones, 2012, p. 10). Such an 'architecture of pleasure' helps to reassure visitors that they are 'in the right place' (Ayala, 1991, p. 569; King, 1997).

The resort tourism landscape is then modified to meet the visual needs of the guests, to conform to the 'myth' of the 'tropical island' (Baldacchino, 2005, 2012; Jaworska, 2017; Lowenthal, 2007) with its elementary units of perception, or 'iconemes' (Turri, 1998). The beach, the most easily recognizable 'iconeme' of the 'tropical island' (dell'Agnese, 2016a), must be white and perfectly clean, without any rubbish brought in by the sea (a very frequent occurrence, even in the Maldives) and without dead leaves, grass, or seaweed. Thus, 'the beautiful beaches that are proudly advertised for all resort islands need comprehensive beach management in order to maintain their attractiveness' (Domroes, 2001, p. 127). The sea, the second 'iconeme,' must be blue and crystal clear; this sometimes involves removing the sea grass that grows in the lagoon, because 'it could limit the snorkeling area, mooring space, and bring sea grass debris onto the beaches which can be aesthetically unattractive to tourists' (Hassan, 2005, p. 44). Even coconut palms, the third 'iconeme' of the 'tropical island,' are altered from their natural state as coconuts are removed to eliminate possible danger for tourists. As remarked by the *Third Maldives Tourism Master Plan* (Ministry of Tourism Arts and Culture, 2007, p. 48), 'the Maldivian tourism product is defined by beautiful tropical islands with palm fringed white beaches and turquoise clear lagoons and its marine environment.' The beautification process also includes the landscaping of green spaces and the 'cuteification' of the fauna. Gardens are often enriched with non-native elements that offer a generically exotic image of the 'Orient,' such as fountains with Japanese carp and water lilies or bougainvillea and orchids.

Architecturally, the first resorts were very simple, with essential structures and small huts with straw roofs and coral walls. They respected the image of 'Robinson for a day,' which a holiday on an otherwise uninhabited 'tropical island' could offer. At the time, it was easy to observe that 'much thought was given to and care was taken to preserve the atmosphere of the desert islands that tourists seek' (Lateef et al., 1980, p. 47). Such simplicity did not last long. The rush toward higher standards of accommodation was stimulated by the intense competition between resorts (Purandare, 2009, p. 7) and by the trend of providing services increasingly at odds with their location. Today, Maldivian resort islands include some of the most expensive holiday resorts in the world.

The frontline and guest relations staff share these spaces with the guests and are part of the dreamscape. Therefore, to avoid altering the myth of the tropical island where everything is pristine and happy, the frontline staff must pretend to have fun even when they are actually working, as Minca writes (2009). Then, there are the Maldivian room boys and waiters. These staff members are also part of that same landscape, although in a subordinate position. For this reason, they are rewarded with tips that, at the end of the month, may sometimes exceed the amount of their salary. In a certain type of resort, there is also the figure of 'Mr. Friday,' a butler at the service of the customer, whose presence emphasizes the 'Robinson island' image. However, the resort, of course, is not a 'desert island'; it is a modern hotel structure that requires energy, water and waste management, repairs, laundry, and kitchens to function. The spaces dedicated to this sort of operation are not part of the dreamscape; they are too banal and contemporary. Thus, they are located within the island's interior, hidden from the tourists' sight, along with those who work to make them operate, i.e. the South Asian backstage workers.

7. The Maldivian resort as a segregated space

As a tourism enclave, the Maldivian resort island is a separate space where only paying guests and international workers are allowed to set foot. Despite this fact, walls or other human-made barriers do not enclose the enclave. Due to its physical structure, the separateness from the broader society of the resort island could be perceived as a consequence of its insularity. However, the external border of the resort island is not a natural border but a social construct—the product of a strategy aimed at maintaining a separation between international tourists, with their expectations and cultural behavior, and the local population, with their social habits and religious norms.

The borderization of the island involves more than just its shoreline. An equally rigid compartmentalization is imposed on the interior, dividing backstage and workers' areas from the tourism spaces of the outer edge. Workers' quarters and service areas are fenced off and made invisible to guests by physical barriers (typically, walls covered with tropical climbing plants to make them look more 'natural'). Tourists are not allowed into the workers' quarters and dormitories, in the staff mosque, or in their recreational spaces. Neither can tourists enter the utility areas where energy is generated and waste and water are treated. Similarly, workers can go only where their work takes them, so their freedom of movement on the island is further limited.

Other borders are less visible but just as powerful. Specific rules regarding the spatial behavior of different categories of individuals also apply in tourist areas. All tourists are allowed to wear swimsuits on the beach, but no one is allowed to wear a swimsuit at the restaurant or outside the beach. Tourists of both sexes can enjoy a spa in a state of almost total nudity, but no worker can enter the spa except for beauticians, who are usually of Balinese or Thai origin. With the arrival of clients from Muslim countries following the expansion of the tourist market, the compartmentalization of tourist spaces has become even more complicated. For these new arrivals, the habit of serving non-halal food must be reverted, and new private areas must be reserved on an island to provide Muslim guests with the freedoms they desire on their holiday. Although Muslim guests may share certain spaces with other guests, such as the restaurants and the lobby, they must book villas with walled gardens and private swimming pools, so that the Muslim women can bathe and enjoy the sun in the exclusive company of their families. In this way, the spatial structure of the resort island is carved up into a complex trans-scalar system: small enclaves (where Islamic religious norms are practiced) are nested within the overall tourism enclave, which, in a sort of set of Chinese boxes (dell'Agnese, 2016b), represents a separate space of non-Islamic norms inside a Muslim country.

In addition to these micro-islands created to guarantee the privacy of Muslim tourists, there are other islands within the island where backstage workers live and work. Differences in religion and place of origin also indirectly affect the sizes of these spaces. This is a distinction operating both in terms of economic opportunities (in addition to wages, there is a big difference in the possibility of getting a tip, for example) and in terms of personal mobility. Expatriates working as frontline staff can usually go to a special portion of the beach in their free time, whereas workers from the Maldives and those from South Asia cannot. Expatriate workers may also share the restaurants and lobby with the guests, but the backstage workers cannot. The accommodations for these two categories of workers are also very different: members of the frontline staff usually sleep in bungalows, close to those of tourists, while backstage workers have their own accommodations on the island in the so-called staff area. Moreover, the small size of the island limits physical space. So 'sharing up to 10 staff members (sometimes more) in a room is common and privacy is an exception' (Ministry of Tourism Arts and Culture, 2011, p. 14). For most of these workers, freedom of movement is strictly limited to the distance between their workplace and their resting place, a range that, in the limited amount of space on the resort island, can be no more than 150 m. Their 'islands within islands' are clearly quite small.

8. Networking the resort?

The massive presence of foreign workers in the tourism sector has been perceived as a problem since the *Second Tourism Master Plan 1996–2005*, which put forward a list of measures to encourage Maldivian workers, both males and females, to obtain jobs in tourism facilities. One of the suggested steps included the possibility of creating part-time jobs in tourist resorts and strengthening transportation networks between tourist resorts and residential islands. Another suggestion stressed the importance of improving accommodation facilities and working conditions in the resorts to increase their attractiveness to Maldivian workers. Such measures would also need to partially address the issue of leakage: 'close to 50% of resort employees are expatriates who repatriate a high proportion of their income.' From an economic point of view, these attempts at connecting the tourism industry (and specifically the resort islands) with the local population were not highly successful. Hence, the *Third Tourism Master Plan*, launched in 2007, marked a policy change, with the aim of 'taking tourism to the people within the broader framework of sustainable development in the economic, environmental, and social spectra' (Kundur, 2012, p. 4). The principal goal of this plan was to 'expand the industry across the country to generate more benefits for the Maldivian people, reduce economic leakage from the tourism sector, and increase employment opportunities for Maldivians by improving human resource development and training for the industry' (Ministry of Tourism and Civil Aviation, 2007, p. 5). To generate additional benefits for the local population, to increase the involvement of the local population in tourism activities and to promote low- and medium-budget tourism revenues, the TTM lifted the ban on renting rooms on islands other than Malé. In addition, there were suggestions aimed at improving the presence of Maldivian workers in the resorts. For example, there was the idea of positioning resorts closer to inhabited islands to facilitate the daily commuting of Maldivian workers and the presence of women workers in the resorts. Other steps were aimed at easing connections between the resorts and the local community, encouraging excursions to the inhabited islands, enhancing the opening of tourist shops and restaurants, promoting cultural and historical sites, and helping resort-community partnership programs and projects. The *Fourth Master Plan, 2012–2017* was permeated by a similar attitude and pushed forward the objective of increasing the territorial integration of tourist facilities, while simultaneously recognizing the environmental issues raised by tourism in the local islands. Hence, to 'assist inhabited islands near resorts to establish a self-sustaining waste management system,' the plan made a suggestion to 'develop waste management plans for inhabited islands in atolls with tourism development with the assistance of tourism facilities in the area' (Ministry of Tourism Arts and Culture, 2013, p. 33). As a real novelty, the plan also tried to promote collaboration between not only resorts and local islands but also resorts and guesthouses. Moreover, this master plan recognized the role of resorts in properly managing the country's reefs (Moritz et al., 2017) and suggested that the tourism industry could be encouraged to play an active role in environmental management and conservation in the Maldives. The objective was to promote the connection between the resorts and their well-managed reefs and the rest of the environment in order to create a 'biosphere reserve.' The general idea was to 'drive the "biosphere reserve" program by setting up marine reserves with in house reefs or boundaries of resorts' (Ministry of Tourism Art and Culture, 2013, p. 32).

Based on these stimuli, or through their initiatives, some resorts have also adopted an open attitude toward the local community and launched projects directed at nearby inhabited islands. For example, *Soneva Fushi* is a resort owned by a corporation listed as one of the pioneers of 'sustainable luxury' because its declared aim is to be 'at the service of our local and wider communities as well as our guests and shareholders' (Gardetti & Girón, 2016). From this perspective, in 2014, the resort launched a 'Learn to Swim' program for the children of the neighboring island of Eydafushi. *Six Senses Laamu,* another resort that once belonged to the same corporation, has developed a series of initiatives to support local schools and a program to strengthen relations with communities on the 11 islands of the Laamu Atoll. The project is aimed at conducting educational sessions with students and developing a platform for sharing ideas on sustainable development with the local population. *Angsana Velavaru*, in the Faafu Atoll, has started a collaboration with the island of Magoodhoo to help the local community with waste management. Some other resorts, such as *aaaVeee* in the

Dhaalu Atoll, have begun to collaborate with guesthouses on nearby islands, opening their beaches to their guests or offering special packages for Maldivian guests and visitors with a business visa. All of these initiatives, even if they are not, at the moment, systemic, can be useful in reducing the separation between the 'tourist islands' and the Maldivian population. Therefore, such initiatives have the ability to promote the territorial integration of resort islands. Even so, these initiatives have no practical effect on the segregation of workers from South Asia living within the resorts.

9. Conclusion: the resort island as an integrated/segregated space

The Maldivian resort island is a 'space apart' from the rest of the sociospatial environment of the archipelago. Separated from the spaces inhabited by the Maldivian population, resort islands are not permanently settled by a resident population but rather are used by only the transient populations of workers and tourists. In their landscape, these islands mirror a spatial ideal—the myth of the 'tropical island' as a 'paradise.' On the island, the image is obtained as the result of a sophisticated artifice and constant maintenance. The limited dimension of the Maldivian coral islands enhances the oxymoronic experience offered by the resort as a 'counter-site,' where it is possible to have everything when nothing is, in fact, easy to acquire.

The 'dreamscape' of the resort island also involves the members of the frontline staff, who are supposed to appear as if they are having fun, while they are actually working (Minca, 2009). In the Maldivian resort islands, the frontline staff are usually expatriates of European origin or are from the Maldives. The room boys and waiters are usually from the Maldives as well, also belonging to the 'dreamscape,' albeit in a subordinate position. In addition, because room boys and waiters are in contact with the public, they enjoy the possibility of earning tips. The third group of workers, however, does not belong to the same landscape. The third group constitutes the laborers working in the maintenance areas, in the kitchens, and in the laundry, that is, in all those areas that are fundamental to the functionality of the resort but are too banal and modern to be shown to the guests. These workers, usually of South Asian origin, are hidden from the view of the tourists, almost 'caged' inside the interior of the resort island. For these workers, the working area and the staff area represent 'islands within islands,' as suggested by Mountz (2014).

The latest Tourist Master Plans have tried in different ways to improve the connection between tourist facilities and local communities. Some resorts, with the aim of improving their overall sustainability performance, have also undertaken collaboration and networking projects with neighboring islands. Such attempts have the potential to improve the territorial integration within these islands. Unfortunately, these programs do not seem to have any influence on the spatial segregation of the expatriate workforce of South Asian origin living inside the resorts. As Baldacchino recalls, 'paradises, but also the Gulags, are generally islands' (2005, pp. 247–248). The same applies to most of the Maldivian resort islands, at least for a significant number of their workers.

Disclosure statement

The authors have no conflicts of interest to declare.

References

Aili, L., Jiaming, L., & Min, L. (2007). Progress in enclave tourism study of overseas: A literature review. *Chinese Journal of Population Resources and Environment*, *5*(3), 76–81.

Ayala, H. (1991). Resort hotel landscape as an international megatrend. *Annals of Tourism Research*, *18*(4), 568–587.

Baldacchino, G. (2005). Editorial: Islands—Objects of representation. *Geografiska Annaler: Series B, Human Geography*, *87*(4), 247–251.

Baldacchino, G. (2012). The lure of the island: A spatial analysis of power relations. *Journal of Marine and Island Cultures*, *1*(2), 55–62.

Baum, T. (1997). The fascination of islands: A tourist perspective. In D. G. Lockhart & D. Drakakis-Smith (Eds.), *Island tourism: Trends and prospects* (pp. 21–35). London: Pinter.

Britton, S. G. (1982). The political economy of tourism in the third world. *Annals of Tourism Research*, *9*(3), 331–358.

Butler, R. W. (1993). Tourism development in small islands: Past influences and future directions. In D. G. Lockhart, D. Drakakis-Smith, & J. A. Schembri (Eds.), *The development process in small island states* (pp. 71–91). London: Routledge.

Carlisle, S., & Jones, E. (2012). The beach enclave: A landscape of power. *Tourism Management Perspectives*, *1*, 9–16.

Cohen, E., & Neal, M. (2012). A middle eastern muslim tourist enclave in Bangkok. *Tourism Geographies*, *14*(4), 570–598.

Connell, J. (2003). Island dreaming: The contemplation of Polynesian paradise. *Journal of Historical Geography*, *29*(4), 554–581.

Deckard, S. (2010). *Paradise discourse, imperialism, and globalization: Exploiting Eden*. London and New York: Routledge.

dell'Agnese, E. (2016a). Tropical island: The power of a landscape myth. Paper presented at the International Geographical Union Joint Symposium of Commissions on Islands, Political Geography, and Geomorphology & Society, NTU Global Change Research Center, Taiwan.

dell'Agnese, E. (2016b). Resorting the island in the Maldives: From the Bali Syndrome to the Kurumba Game. Paper presented at the 33rd International Geographical Congress, International Geographical Union, Beijing.

Domroes, M. (2001). Conceptualising state-controlled resort islands for an environment-friendly development of tourism: The Maldivian experience. *Singapore Journal of Tropical Geography*, *22*(2), 122–137.

Edmond, R. (1997). *Representing the South Pacific. Colonial discourse from cook to Gauguin*. Cambridge: Cambridge University Press.

Firaag, I. (1997). Tourism and the environment: Current issues for management. In D. J. Nickerson & M. H. Maniku (Eds.), *Workshop on integrated reef resources management in the Maldives (pp. 187–198)*. Madras, India: Bay of Bengal Programme.

Gardetti, M. A., & Girón, M. E. (2016). *Sustainability luxury and social entrepreneurship II: More stories from the pioneers. Sheffield*, UK: Greenleaf Publishing.

Hassan, M. Z. (2005). *Environmental guidelines for tourist resort development & operation in the Maldives: A guide for planners, managers & operators*. Maldives: Ministry of Tourism.

Jaworska, S. (2017). Metaphors we travel by: A corpus-assisted study of metaphors in promotional tourism discourse. *Metaphor and Symbol*, *32*(3), 161–177.

King, B. E. M. (1997). *Creating island resorts*. London: Routledge.

King, B. E. M. (2001). Resort-based tourism on the pleasure periphery. In D. Harrison (Ed.), *Tourism and the less developed world: Issues and case studies* (pp. 175–190). Wallingford: CABI Publishing.

Krippendorf, J. C. (1987). *The holidaymakers*. London: Heinemann.

Kundur, S. K. (2012). Development of tourism in Maldives. *International Journal of Scientific and Research Publications*, *2*, 223–228.

Kundur, S. K., & Murthy, K. (2013). Environmental impacts of tourism and management in Maldives. *International Journal of Environmental Sciences*, *2*(1), 44–50.

Lateef, K. S., Loos, J., Santen, G. V., Doggart, C., & Tanaka, Y. (1980). *The Maldives: An introductory economic report*. Washington, DC: World Bank Country Study.

Lea, J. (1988). *Tourism and development in the third World*. London: Routledge.

Lowenthal, D. (2007). Islands, Lovers, and Other. *Geographical Review*, 97, 202–229. doi: 10.1111/j.1931-0846.2007.tb00399.x

MacCannell, D. (1992). *Empty meeting grounds: The tourist papers*. London: Routledge.

Minca, C. (2000). 'The Bali Syndrome': The explosion and implosion of 'exotic' tourist spaces. *Tourism Geographies*, *2*(4), 389–403.

Minca, C. (2009). The island: Work, tourism and the Biopolitical. *Tourist Studies*, *9*(2), 88–108.

Ministry of Tourism. (1996). *Review and recommendations Maldives Tourism Master Plan 1996 – 2005*. Male': Ministry of Tourism.

Ministry of Tourism. (2016). *Tourism yearbook 2016*. Male': Republic of Maldives.

Ministry of Tourism. (2017a). *Maldives visitor survey*. Male': Republic of Maldives.

Ministry of Tourism. (2017b). *Tourism yearbook 2017*. Male': Republic of Maldives.

Ministry of Tourism and Civil Aviation. (2007). *Maldives Third Tourism Master Plan 2007-2011*. Male': Ministry of Tourism and Civil Aviation.

Ministry of Tourism Arts and Culture. (2011). *Strategic human resource development plan for the tourism industry 2011 – 2015*. Fourth Draft, Male': Republic of Maldives.

Ministry of Tourism Arts and Culture. (2013). *Fourth Tourism Master Plan, 2013-2017*. Male': Republic of Maldives.

Mohamed, A. Z. (2013). The expatriate workforce: Boon or bane? *Quarterly Economic Bulletin*, *13*, 115–124.

Moritz, C., Ducarme, F., Sweet, M. J., Fox, M. D., Zgliczynski, B., Ibrahim, N., … Abdulla, A. (2017). The "resort effect": Can tourist islands act as refuges for coral reef species? *Diversity and Distributions*, *23*(11), 1301–1312.

Mountz, A. (2014). Political geography II: Islands and archipelagos. *Progress in Human Geography*, *39*(5), 636–646.

National Bureau of Statistics. (2014). Maldives population and housing census 2014, release IV: Employment. Retrieved from http://statisticsmaldives.gov.mv/statistical-release-iv-employment/

National Bureau of Statistics. (2017). *Statistical pocketbook of Maldives 2017*. Republic of Maldives: Ministry of Finance & Treasury Male'.

Niyaz, A. (2017). *Maldives tourism: A historical perspective*. Male': MMPRC.

Price, R. G. A., & Firaq, I. (1996). The environmental status of reefs on Maldivian resort islands: A preliminary assessment for tourism planning. *Aquatic Conservation: Marine and Freshwater Ecosystems*, *6*(2), 93–106.

Purandare, J. (2009). How can tourist developments in popular tourist destinations, such as the Maldives, *also be sustainable developments? International Journal of Small Economies*, *1*(1), 1–14.

Rufin-Soler, C. (2005). Les politiques de gestion insulaire dans l'archipel des Maldives. *Cybergeo: European Journal of Geography*, 322, 17, doi: 10.4000/cybergeo.3067.

Saarinen, J. (2016). Enclavic tourism spaces: Territorialization and bordering in tourism destination development and planning. *Tourism Geographies*, *19*(3), 425–437.

Saeed, S., & Annandale, D. (1999). Tourism and the management of environmental impacts in the republic of Maldives. *Journal of Policy Studies*, *7*, 81–88.

Sathiendrakumar, R., & Tisdell, C. (1989). Tourism and the economic development of the Maldives. *Annals of Tourism Research*, *16*(2), 254–269.

Scheyvens, R. (2011). The challenge of sustainable tourism development in the Maldives: Understanding the social and political dimensions of sustainability. *Asia Pacific Viewpoint*, *52*(2), 148–164.

Scheyvens, R., & Momsen, J. (2008). Tourism in small island states: From vulnerability to strengths. *Journal of Sustainable Tourism*, *16*(5), 491–510.

Shakeela, A., & Cooper, C. (2009). Human resource issues in a small island setting the case of the Maldivian tourism industry. *Tourism Recreation Research*, *34*(1), 67–78.

Shakeela, A., Ruhanen, L., & Breakey, N. (2011). The Maldives: A sustainable tourism success story? *e-Review of Tourism Research*, *9*(5), 243–264.

Shakeela, A., & Weaver, D. (2017). The Maldives: Parallel paths of conventional and alternative tourism. In C. M. Hall & S. Page (Eds.), *Routledge handbook on tourism in Asia* (pp. 265–274). London: Routledge.

Shaw, G., & Agarwal, S. (2012). Changing geographies of coastal resorts: Development processes and tourism spaces. In J. Wilson (Ed.), *The Routledge handbook of tourism geographies* (pp. 240–248). London: Routledge.

Sobir, A. R., Shiuna, F., Ibrahim, L., & Shafeeq, S. (2014). *Maldives human development report 2014. Bridging the divide: Addressing vulnerability, reducing inequality*. Male': Ministry of Finance and Treasury and the United Nations Development Programme in the Maldives.

The World Bank. (2013). The Maldives: A development success story. Retrieved from www.worldbank.org/en/results/2013/04/10/maldives-development-success-story.

Turri, E. (1998). *Il paesaggio come teatro [Landscape as a theatre]*. Padua: Marsilio.

UNDAF. (2015). *United Nations development assistance framework 2016-2020*. Male: Government of Maldives and the United Nations in the Maldives.

World Travel & Tourism Council. (2015). *WTTC travel & tourism economic impact. Maldives*. London: WTTC.

Zubair, F. N. I., & Bouchon, F. (2014). Maldives as a backpacker's destination: Supply and demand perspectives. *Procedia - Social and Behavioral Sciences, 144*, 256–263.

Does geography matter in all-inclusive resort tourism? Marketing approaches of Scandinavian tour operators

Sandra Wall-Reinius, Dimitri Ioannides and Kristina Zampoukos

ABSTRACT
Over the last decade, there has been noticeable rise in popularity of all-inclusive holidays. This growth has coincided with the propensity in many destinations to develop tourism enclaves, which can either be purpose-built gated resorts physically isolated from their surrounding community or appear in the form of cruises, which have emerged as a particularly popular form of travel. In this explorative paper, we focus on the marketing of all-inclusive holidays, specifically those occurring in enclaves (including cruise ships). We investigate to what extent the geographic location of the tourist enclave is an important consideration for the travel industry. In other words, when it comes to all-inclusive holiday products, do the place-based attributes on offer at the destination and the actual location of the holiday matter from the perspective of those who are creating and selling the travel packet? An explorative study of Scandinavian tour operators shows that the local setting of the holiday is in fact a secondary consideration compared to the services and facilities on offer. Thus, there is an overriding tendency to downplay the destination's place-based attributes and it does not seem so important where the all-inclusive resort is located as long as it is well connected to the market and promises a comfortable holiday to the consumer. Tourism enclaves in the context of placelessness are discussed.

Introduction

In *The Guardian*, on 8 March 2014, reporter Tracy McVeigh discussed the growing popularity of all-inclusive holidays for consumers who prefer packaged forms of travel. She indicated that many British tour operators now almost exclusively offer this type of holiday. McVeigh's argument was that the growth in this travel product where one pays upfront for everything and is encouraged to 'leave the wallet at home' makes particular sense to budget-conscious consumers. Meanwhile, *Travel Weekly* (Myers & Jainchill, 2015) recently mentioned that the expansion of all-inclusive offerings in the Caribbean and Mexico is no longer a phenomenon restricted to families on a tight budget. Rather, this form of travel now appeals to persons of varied financial means including the wealthy who demand to be spoiled by an all-expenses-paid holiday in a luxury resort.

All-inclusive holidays go hand in hand with the development of purpose-built resort enclaves that are physically segregated from the host community (e.g. Domroes, 2001; Meyer-Arendt, 1992; Saarinen, 2017). From the perspective of the tourists who demand these holidays this form of stand-alone, highly regulated development is attractive since it removes the element of dealing with unexpected surprises in an unknown environment. Especially for tourists who wish to have a carefree and relaxing holiday, there is a high element of comfort associated with visiting a predictable space where everything is taken care of by someone else. Today, all-inclusive resorts have become increasingly popular worldwide often resulting from destination policies favouring foreign direct investment, which in turn are reflected in large-scale, planned projects controlled and managed either by powerful local elites or major multinational concerns (e.g. Manuel-Navarrete, 2016; Sidaway, 2007). The proliferation of these forms of development is yet another expression of what Harvey (2013) describes as capitalism's constant obsession with profit in an era of neoliberalism (see also Mosedale, 2016), something we also notice in the constant redevelopment areas of urban cores of established metropolitan regions and the increasing popularity of brand new leisure-oriented cities in countries like the United Arab Emirates, China, and South Korea (Bagaeen, 2007).

All-inclusive enclaves are popular in coastal regions in countries like Turkey, Egypt, Mexico, Jamaica, and other parts of the Caribbean, but also the Indian Ocean. The famed Jamaican resorts under the Super Clubs label (Issa & Jayawardena, 2003) as well as the many stand-alone – once uninhabited – island resorts found in the Maldives (Domroes, 2001) are particularly notable examples. To be sure, the facilities and activities offered vary from place to place. Nevertheless, typically in addition to accommodation one finds sports facilities (e.g. tennis, golf, and fitness centres), recreational amenities (e.g. pools and spas), activities for children, and a multitude of eating and drinking places embraced by modern 'international' style and standard.

Researchers have engaged in many discussions and developed extensive conceptualizations leading to multiple and complex meanings of place (e.g. Creswell, 2004; Edensor, 2007; Massey, 1991, 2005). Places are 'spaces which people have made meaningful' (Cresswell, 2004, p. 7). Nevertheless, when it comes to tourism-specific enclaves, these often emerge as an outgrowth of ubiquitous forces of globalization, mass commercialization, and communication that have, to various degrees, homogenized geographical areas into standardized destinations where they look and feel alike, offering uniform products and services designed to generate similar experiences (see also Birkeland, 2008). This anonymization and homogenization renders such spaces 'placeless' meaning that the significance of local-based attributes is underplayed and minimal connections exist with the local culture (Arefi, 2007; Entrikin, 1991; Freestone & Liu, 2016; Relph, 1976).

Our use of the term 'placeless' draws from Relph (1976) who first coined the concept in relation to modern urban environments, constructed without clear consideration of the significance of place. His argument was that this phenomenon results in the erosion of the distinctiveness of place and in the making of standardized landscapes. Thus, '"Placelessness", ... is a weakening of the identity of places to the point where they not only look alike but feel alike and offer the same bland possibilities for experience' (Relph, 1976, p. 90). Referring specifically to tourism architecture and landscapes, Relph indicates that the tourism sector emerges as 'an homogenizing influence' (p. 93), since it targets outsiders, spectators, and consumers. The result is 'synthetic landscapes' and 'pseudo-places' (p. 93) encouraging an uncritical acceptance of mass values. Other studies of placelessness – or nowhereness – in tourism, focus on airports (Rowley & Slack, 1999), shopping malls (Shim & Santos, 2014), festivals (MacLeod, 2006), and amusement parks (Warren, 1999). They recognize that these places or venues are also established without regard to local characteristics. Rather, they frequently represent a modernized, illusionary, and enclosed commercial space (see also Harvey, 2013). Placelessness is also reflected in cruise tourism, perhaps the ultimate representation of all-inclusive holidays. One of the products offered by Carnival Cruises are 'cruises to nowhere' advertised in this manner: 'Where is the best place for you to spend a few fun-filled days? Nowhere! A Carnival 'Cruise-To-Nowhere' is the perfect way to get away for a couple of days' (Carnival Website, 2014).

Based on current trends in the travel industry, it appears the concept of enclave-style tourism in the mould of cruises or all-inclusive land-based resorts is here to stay for the predictable future. The launching of *The Harmony of the Seas* (Boyle, 2016) in early 2016 with a capacity to accommodate almost 7000 guests and the development or renovation of several all-inclusive resorts in many parts of the world (Reynolds, 2015) certainly suggest the confidence by many in the tourism industry that the demand for this type of holiday is strong and, indeed, on the rise. The increasing occurrence of enclosed spaces (all-inclusive resorts and cruises) directly contradicts other trends in the tourism sector, such as cultural tourism, geotourism, ecotourism, and ethnic tourism, whereby the importance of place and unique experiences of local heritage, culture, cuisine, nature, and atmosphere emerge as key attractions for tourists (Mattsson, 2010). The nagging questions arising in relation to all-inclusive developments are: what does this tell us about the role of place and the (non)importance of place in contemporary tourism? Do destinations' place-based attributes matter or have the geographical features become secondary to other factors such as accommodation facilities, activities on offer, or value for money? How important is the actual geographic location of an all-inclusive enclave to those who market the destination and how are all-inclusive holidays promoted? The overall aim is to explore the role a geographical location plays in contemporary mass tourism by focusing on all-inclusive developments. Hopefully, this paper will contribute to the discussion about how tourism shapes spaces and their emerging interconnections, while also identifying the influence tour operators (as key gatekeepers of tourism development) play in promoting that form of mass tourism.

The paper unfolds as follows. First, we review work addressing enclave tourism and all-inclusive holidays. We briefly examine the forces leading to the proliferation of this type of tourism and, thereafter, we present sources of data and findings of a study targeting Scandinavian tour operators. We end by discussing the role of local features and analysing the continuum of enclavic tourist spaces and placelessness.

Enclavic tourist spaces

Discussions relating to enclavic spaces are popular in human geography and related subjects. In *Rebel Cities* (2013), David Harvey underlines how in the spirit of rampant neoliberalism, parts of cities are constantly transformed into privatized or semi-privatized spaces patrolled by security officers and monitored by surveillance equipment, promising those with the financial means the ability to safely pursue their desired consumer-led lifestyle. Likewise, Judd (1999) refers to the 'tourist bubble' to illustrate the reshaping of contemporary inner cities (especially in North America and Western Europe), whereby signature waterfront redevelopments, convention centres, and festival market spaces stand isolated from surrounding neighbourhoods and their inhabitants. He argues that standardized international design and well-known chain retail outlets, eateries, and hotels dominate these spaces. These characteristics together with regularly staged spectacles and performances project to the average visitor a high sense of predictability. On a grander level, mega tourist bubbles have appeared in the form of planned new towns geared towards tourism and leisure (Foglesong, 2001). In East and South Asia, for example, tourism/leisure cities of varying scales have become enormously popular in the last decade (Douglass & Huang, 2007).

The concept of spatial bubbles is hardly new nor solely urban-specific. Cohen (1972) first discussed the 'environmental bubble' relative to tourism while Edensor (1998) depicted organized tourist spaces as enclavic. Edensor viewed these spaces, which were either foreign-controlled or run by local elites, as privatized, pre-planned, staged, commoditized, and regulated, but also bounded and isolated from their surroundings. He explained that typically their international style starkly contrasts to the surroundings. Boundaries separate and regulate encounters between the familiar, safe, and comfortable spaces for tourists and the exclusionary and exploitative spaces for workers and surrounding local population (Edensor, 2007; Manuel-Navarrete, 2016). Additionally, such spaces can exacerbate destinations' dependency on metropolitan countries and transnational tourism companies (Aili, Jiaming, & Min, 2007; Andersen, 2011; Britton, 1979, 1982; Mbaiwa, 2005; Shaw & Shaw, 1999). Shaw and Shaw claim that tourism enclaves reinforce neocolonial patterns of socioeconomic and spatial polarization, while Bunnell, Muzaini, and Sidaway (2006) and Manuel-Navarrete (2016) discuss how enclaves controlled by government-backed corporate actors complicate access for locals to land and resources (see also Bladh & Holm, 2013; Lee, Hampton, & Jeyacheya, 2015). Further, enclavic tourism generally displays weak linkages with existing economic sectors (Freitag, 1994) and leads to high monetary leakages (Aili et al., 2007), thus becoming economically unsustainable. In enclave tourism, tourists are motivated to stay within the resort, limiting their movement as well as their spending (Andersen, 2011; see also Meyer-Arendt, 1992).

Cruise ships also come under the purview of tourism enclaves. 'A cruise ship is a controlled, safe, pleasurable environment with a wide range of recreational facilities and activities. This as a tourist bubble is a place designed exclusively for tourists and those that serve them' (Jaakson, 2004, p. 46). Cruise ships are 'spaces of containment' (Weaver, 2005, p. 165), floating travel destinations where tourists spend all or almost all their time for the duration of the trip. Wood (2000) views them as examples of 'economic and political deterritorialization' (p. 358) given their freedom from various regulations related to land-based sectors and their ability to employ workers from countries all over the world. Several cruise companies shun established destinations, spending the entire trip at sea or visiting one of their privately owned or leased islands or beaches (Gui & Russo, 2011; Rodrigue & Notteboom, 2013) such as the heavily patrolled Labadee beach on Haiti owned by Royal Caribbean International (Royal Caribbean International, n.d.). This allows them to maximize their revenue capture (Weaver, 2005).

All-inclusive tourism

While the cruise ship is the quintessential example of 'deterritorialization' (Wood, 2000, p. 345) on the sea, its direct contemporary land-based equivalent is the all-inclusive resort. Indeed, both cruises and all-inclusive operations have emerged as significant niches and fast-growing subsectors in global tourism (Issa & Jayawardena, 2003; Rodrigue & Notteboom, 2013). All-inclusive holidays include everything in a pre-paid price, eliminating those troublesome extra charges incurred on holiday. Thus, the holiday becomes hassle-free and the tourist enjoys meals, beverages, including alcoholic drinks, and other services and activities without spending anything extra. Various observers conclude that all-inclusive tourism is particularly attractive for visits to distant counties with different cultures, unreliable transportation and uncertain hygiene, safety, and security conditions (Andersen, 2011; Wong & Kwong, 2004).

Regardless, an exact definition for the term 'all-inclusive' is elusive since there exists inconsistency as to which components the all-inclusive holiday package might include. Often, the transfers to and from the airport are included in the all-inclusive package (Conrad, 2011). When visiting Club Med in Cancun, Mexico, the pre-paid package includes transfers, accommodation, meals, drinks, and various activities, but excludes high-end drinks like champagne as well as the fees for some activities. Issa who in 1976 opened Negril Beach Village (later known as Hedonism II) as a property for young singles has claimed credit for the original enclave-style all-inclusive destination. Despite encountering considerable scepticism, Issa was convinced his idea would work because he noted during the 1970s' oil crisis that Club-Med and the cruise industry, which had operated on a similar basis, weathered the economic crisis better than other types of properties (Issa & Jayawardena, 2003).

Prior to 'all-inclusive' becoming specifically associated with enclave resorts, it was used to describe package holidays tailored by tour operators and had become a key reason behind mass tourism's phenomenal rise in the 1960s. The tour operator achieves significant economies of scale by bundling into a package the individual components contributing to the total traveller experience (Ioannides, 1998). Recently, the inclusive package tourism market, especially in Europe, has become increasingly controlled by a small number of large companies many of which grew through horizontal and vertical integration (Koutoulas, 2006). Thus, faced with constantly expanding capacity, the locally owned accommodation facilities in various destinations are at an increasingly weak bargaining position *vis à vis* these powerful exogenous players when negotiating prices. Although the tourism industry remains characterized by small businesses at destination level, it is evident that the increasing power of multinational tourism groups follow strategies of global expansion, and hence, local producers interrelate more and more with external players of international tourism chains.

Tour operators, as one example of the international tourism industry, notoriously display minimal loyalty to specific destinations (or businesses within destinations) preferring instead a multi-locational approach that targets holiday type and not a specific geographic region (Ioannides & Debbage, 1998; Lee et al., 2015). In the discourse of place competitiveness, places have become increasingly interchangeable and at the same time spatial homogenization and standardization are by-products and processes of rational economic planning, consumerism, and transnational corporations that characterize modernity, globalization, and the neoliberal economy. Contemporary capitalism and the neoliberal economy leads to foreign capital investments in large-scale all-inclusive resort developments whilst also causing dependencies and risks associated with exogenous factors tied to the wider global system (Lee et al., 2015). The production of enclavic tourist spaces, as all-inclusive seaside resorts and cruises, is of course part of wider processes and in relation to this it is relevant to discuss enclosed, standardized, and anonymous spaces in the context of 'placelessness' as described by Relph (1976) and others. In order to explore tourist enclaves and the evident *non-significance* attributed by key gatekeepers to the locational (place-based) factors of such spaces, we have focused on the manner in which tour operators situated in key origin countries (in this case Scandinavian countries) market all-inclusive holidays.

Method

The concept of all-inclusive holidays has gained increasing popularity in several European countries. A study undertaken for the European Commission (TNS Political & Social, 2015) on Europeans' preferences towards tourism indicates that 30% opted for an all-inclusive trip. The percentage of those between the ages of 15 and 34 on such holidays was higher (38%). There are considerable differences between different markets. For instance, while 39% of Belgians chose all-inclusive holidays, in Sweden, 17% did so. Nevertheless, all-inclusive holidays are especially popular for Swedish families with younger children as reported by the Swedish Chambers of Commerce (2011). When it comes to all-inclusive holidays, tour operators remain a dominant player for developing and promoting the concept, including packaging of all components into one price (Van Wijk & Persoon, 2006). This boils down to their ability to generate significant economies of scale but also their control over quality and their ability to substantially reduce risks while placing client satisfaction high on the agenda.

In this study, we have chosen to examine Scandinavian tour operators (in Denmark, Norway, and Sweden) who focus on all-inclusive travel. In the Scandinavian countries, charter travel has been popular for decades but lately all-inclusive holidays, in particular, have gained increased media attention (for example, in 2008, the journalist Jennie Dielemans wrote the book *Welcome to Paradise* while other critical investigations of the tourist industry were carried out by associations like Schyst Resande - *in English Fair travel* – 2011, 2012). We were first inspired to select the Scandinavian market because of an advertisement in a Swedish magazine about 'Playitas,' an all-inclusive sport and activity resort promoted by the tour operator Apollo. Although the ad included pictures and information on things like sport activities, yoga, nutrition counselling, drinks in the bar, accommodation in apartments, price and departure from the Swedish airport Arlanda, it provided no information whatsoever concerning details such as destination airport, location, or even the country where the resort is located; nor did it mention things for the tourists to do or experience outside the resort. This advertisement inspired us to formulate the theoretical problem and the research questions, and we felt that an investigation of the Scandinavian industry would provide the answers; does geography matter in all-inclusive resort tourism?

Initially, back in April 2011, we conducted an Internet search, using basic Google search for the following keywords: all-inclusive travels; all-inclusive hotels and resorts and; cruises, promoted and sold in Scandinavia. The search yielded 17 different tour operators operating in Denmark, Norway, and Sweden. We conducted a content analysis of images and texts of all these 17 tour operators' websites (accessed April–June 2011). Online content analysis provides an insightful basis for discourse analysis in tourism (Hallett & Kaplan-Weinger, 2010), giving a first image of a destination, and also destination promotion (Law & Chen, 2012). We applied a content analysis of images and texts to explore dominant representational themes in the marketing of all-inclusives. We did this manually; the first step was to search for repeated topics that cropped up in the promotional material, and second, to take notes and write short descriptive summaries of images and promotional text. From the transcripts of collected material, we identified major themes based on analytic induction and comparison of the material (Silverman, 2013).

Because considerable time had lapsed since that initial content analysis, the websites of each company were revisited in October 2016. This allowed us to check whether a given company was still active and if the content of the site has changed significantly. Based on this procedure, out of the 17 Scandinavian tour operators, 2 non-functioning websites were identified (Enjoy Caribbean website and India travel website). A further online investigation suggests that these two companies no longer exist, at least under those names.

Back in April 2011 we also conducted interviews with five Scandinavian tour operators, namely four large-scale operators Fritidsresor/TUI, Apollo, Ving, and Detur, as well as one niche tour operator, India Travel. We asked company representatives to define all-inclusive resorts, and reflect on matters relating to supply and demand. We also asked them to reflect about the trends in the all-inclusive market and to explain how they select the resorts on their itineraries. The interviewees were two communication managers, one senior sales executive, one coordinator for Corporate Social Responsibility, and one vice director. Each interview took about 30 minutes, and was immediately transcribed. Follow-up interviews were carried out in October and November 2016 with three of the five companies. As we have already mentioned, one operator, India Travel, has ceased to operate under that name, and one operator, Fritidsresor/TUI, declined to participate because of reorganization, limited human resources, and lack of time. In the 2016 follow-up interviews, questions were asked to identify changes in supply and demand, if the companies had changed their strategies or focal areas, and what they had to say about the trend of all-inclusive travels. Each interview lasted 30 minutes, was recorded and immediately transcribed. People interviewed included one PR and social media manager, one communication manager, and one sales and marketing coordinator. Once more, the empirical material (websites and interviews) were thematically analyzed where themes and patterns across the data emerged through analytic induction and comparisons (Silverman, 2013).

Findings

First, we tried to gauge how tour operators define all-inclusive holidays, what they focus on, and what they believe their clients desire. The interviewed representatives of the five companies we originally interviewed indicated they offer packages of travel, accommodation, food and beverages, while, sometimes, they also add activities (interviews with representatives from Fritidsresor/TUI, Apollo, Ving, Detur, and India Travel). The amount of food on offer varies. For example, how much food is included in the package depends, according to one interviewee, on the hotel's standard: 'In the simpler hotels with all-inclusive, the food buffet is small. The more luxurious the hotel is, the more food, drinks, and activities are included' (Ving representative, 2011). The representative of Detur indicated the existence of the 'Ultra all-inclusive' package that, in addition to transportation, lodgings, and food, includes beauty/spa treatments and imported beverages.

A fundamental determinant behind their clients' motivation in demanding all-inclusive holidays seems to be the simplicity emerging from bundling various elements of the holiday experience into a single package for a single price. The representative from Apollo pointed out that 'customers want to be comfortable during their holiday and not have to think about the bill.' The representative of Ving claimed that people choose all-inclusive holidays because of the affordability of the experience and because they have the assurance of knowing exactly how much it will cost. Because Ving's clients are mostly families with children they demand the best value money can buy, especially with regards to the availability of food (e.g. meals and snacks catering to children that are available anytime they want these). Another interviewee mentioned that the 'all-inclusive is predictable, comfortable and the costs are controlled.' The idea is, as the representative from India Travels mentioned, that by knowing what is included in the travel package and what is not, surprises are eliminated. Controlling costs seems to be especially important during economic downturns, a point made by the coordinator of Fritidsresor/TUI who argued that 'the latest recession has led to growing demand for all-inclusive holidays.' In the follow-up interviews in 2016 with three of the tour operators, it appeared that there are no changes regarding what their clients demand; all-inclusive travels are considered to be comfortable, simple, and value for money. Additionally, one of the interviewees added: 'there is a sense of luxury and freedom for children, such as being able to get an ice cream or a soda' (Ving representative, 2016). Another of the interviewed tour operators mentioned that fitness and active sport holidays is a growing trend at the moment (Detur representative, 2016).

The importance of a single price is also highlighted on the websites of the 17 companies we investigated for our content analysis. For example, Apollo's website mentions that staying in an all-inclusive hotel:

> does not only mean a lot of good food for the holiday – it is particularly amazingly comfortable and excellent value for money. You eat and drink as much as you want and you do not have to think about the bill. Instead, put your energy on socializing and have fun together with your family on the holiday. (Apollo Website)

The Norwegian company Soli puts the importance of not having to worry about price into a simple perspective by mentioning on their website that an all-inclusive is 'a holiday without a wallet' (Soli Website). From the website analysis it was found that the larger tour operators (e.g. Ving, Apollo, Fritidsresor/TUI, Detur) market the whole concept of all-inclusive holidays and the benefits of having everything included in the price, while smaller companies and niche-oriented travel companies highlight activities within the resort.

That the popularity of all-inclusive holidays, at least for the Scandinavian market, has increased was a trend indicated by four of the five company representatives interviewed in 2011. For example, in 2011, the Ving representative indicated that whereas a decade earlier demand for all-inclusive resort holidays was only between 2% and 3%, now 25% of all trips offered by the company were all-inclusive. Ving has to constantly search for more all-inclusive hotel properties to match the growth in demand. In 2011, Ving, Fritidsresor/TUI and Apollo estimated the share of all-inclusive holidays at about 20%–25% of all their sold products. However, during 2011, Apollo observed a 5%–10% decline because of the situation at the time in Egypt and Tunisia, where they had many all-inclusive hotels (interview with media manager Apollo 2011). In 2011, the interviewed CSR coordinator for Fritidsresor/TUI described that demand had increased dramatically during the period 2005–2010, but that the last year showed a more stable demand. He associated the growth in demand with the widespread recession that was occurring at the time. In the follow-up interviews from 2016, Ving and Apollo could confirm that after a very strong growth of all-inclusive holidays, the trend has stagnated and even decreased slightly. This is because of reduced demand for travel to Egypt and Turkey due to the political instability, and 'now people tend to go Spain or Greece where all-inclusive holiday are not as widespread' (Ving representative, 2016). Regardless, the tour operator Detur reports a growing trend. In 2011, about 50% of their travels were to all-inclusive hotels, while by 2016 they had increased to 70%, and they believe this proportion will keep growing (Detur representative, 2016). In fact, Detur has chosen to increase the offer of all-inclusive holidays to Greece and Spain to compensate for reduced demand of travels to Egypt and Turkey.

According to the Detur representative, an all-inclusive holiday 'is more popular among young people who travel in groups and families with children.' Similarly, the interviewee from Apollo highlighted that all-inclusive trips are 'popular with so-called generation travels, meaning the grandmothers and grandfathers who invite their children and grandchildren' for a family reunion. Even the vice director at the specialist tour operator, India Travels, which only began operating in 2006, noted a growing number of requests for all-inclusive holidays. By 2011, these had become their main product. This interviewee explained that:

> most of those who travel with India Travels have previously travelled to India when they were younger, but then they were backpackers. Now, they want to experience it again but lack the strength/energy to arrange the accommodation on their own and do not want to risk finding a hotel when travelling alone with family.

It was also mentioned by the same respondent that India Travels also has customers who have never been to India before and for them it is especially important from a safety and security point of view to have everything pre-arranged. Further, it seems that the actual demand for all-inclusive resorts has risen because tourists perceive these to be more secure. For instance, one respondent explained the demand for all-inclusive holidays in countries such as the Dominican Republic as having something to do with the fact that 'a Swedish couple was murdered in the 80s, which made tourists afraid. In order to get the tourists back, controlled tourist areas with all-inclusive were constructed.'

Is there a preference on behalf of the tour operators to target certain destinations as opposed to others? A cursory examination of various websites reveals the major companies like Apollo, Ving, and Fritidsresor/TUI serve several destinations. In fact, they tend to offer all-inclusives in the countries that their own charter airlines fly to. By contrast, a travel specialist such as India Travels, Turkietresor, or Detur only offers travel to a single or a narrow range of destinations. In 2011, the interviewed tour operators highlighted Egypt, Tunisia, Turkey, Spain, and some countries in the Caribbean (most notably the Dominican Republic) as the countries with all-inclusive resorts where they choose to offer packages to. There appeared to be a tendency by some of these companies to shun destinations like Greece since all-inclusive properties there were rare. One company representative had this to say on the matter:

> We have had cooperation with Greek hotels for 30-40 years and they are often small family-owned hotels, which do not suit the demand for all-inclusive tourism. However, in Egypt there were no tourist resorts before and no supply of restaurants and activities, so all-inclusive hotels were built and gave the possibility for tourism development. They chose to build [integrated] tourist areas around the nice beaches. (Ving representative, 2011)

Such developments explain the shift in attention towards destinations like Egypt, at least before the more recent political instability in that country.

As to whether or not the operators demonstrate a high level of loyalty to particular destinations and/or hotel properties this is what they had to say. They agree they periodically change destinations. They also choose to switch the hotel property they are collaborating with within the same destination if, for instance, they find another property that offers a similar product for a more competitive price. The reasons for dropping a destination vary and could simply have to do with declining demand. Instead, the companies choose to focus on competing destinations, which are more easily accessible, especially if they allow direct charter flights to and from the Scandinavian countries. An additional issue is that during times of crisis the operators will assess the situation and if they think the risk is too great or that customers will be turned off then they will, at last temporarily, drop the destination.

A major thrust of our research was to identify if the specific geographical location where an all-inclusive resort is situated is highlighted in the respective marketing materials of the tour operators. From the outset it appears that whilst most companies are likely to stress through their websites the benefits of the all-inclusive holiday in terms of the facilities/activities on offer, not to mention the availability of food and beverages, it is often impossible to pinpoint where their properties actually are. The website, for instance, of a company offering all-inclusives to the Caribbean highlights the facilities, the beaches, the Jacuzzis, and other amenities but as far as where the depicted resorts are located is anyone's guess; the happy people on the ubiquitous beach enjoying a sunset could really be anywhere in the Caribbean, or for that matter, on a coast anywhere in the world's middle latitudes (Enjoy Caribbean website). Another tour operator that concentrates on Tunisia and Turkey pays most attention to the sporting facilities located at the respective properties on offer while saying very little on its homepage about the exact localities the properties are situated in (Detur website).

Of course, one may expect certain companies specializing in a narrow range of destinations or ones concentrating on relatively small market niches to be likelier to highlight at least some geographical aspects of their respective destinations. Certainly, one of the companies specializing in all-inclusive tourism to Turkey provides information on its start-up page about Turkish history, geography, and local food (Turkietresor website). It is even possible to obtain general information about particular municipalities, which are located close to the respective all-inclusive hotels on offer. However, when one clicks on the website of one of the all-inclusives on offer, the emphasis is really on the property, the facilities on offer, the availability and range of food and drinks, as well as the various amenities. Even though one can click onto a map indicating where in Turkey the site in question is situated, it is obvious from the content of the website that the information provided is geared towards visitors who rarely if ever wish to leave the premises of the all-inclusive property (Turkietresor website).

The case of the specialist Danish company Nyati Tours that offers safari travel packages to South Africa is of particular interest. One of the main specialties is the all-inclusive Nyati Safari Lodge. The exact location of the lodge is not clear. We are told on the website that it is located 'by Great Kruger National Park in South Africa.' Rather, it is more important to convey to the potential traveller the following:

> The surroundings should be unique, accommodation top-of-the-range, excursions and activities should offer adventure and the culinary aspects should be an experience in their own right. Danish staff permanently resident on-site ensure a safe and well-functioning holiday environment. The all-inclusive principle in South Africa allows guests to enjoy their holiday, without having to worry about unexpected additional costs while there. The wide range of optional activities makes sure to meet individual interests and gives guests the ability to set the pace of their holiday themselves. (Nyati website)

Elsewhere the same website informs the reader that the safari lodge is of the highest quality and although it is built in a rustic 'African style' – no definition what that is – one is assured to find 'construction according to the highest Danish quality standards!' Further, the tourist is able to gaze on the wildlife (e.g. crocodiles and hippos), from her own 'private' terrace while sipping a cocktail. It is also possible to have an 'out of Africa' picnic style breakfast in the middle of the bush something the company labels the 'Blixen Table.' The fact that the Danish author Karen Blixen is associated with Kenya, not South Africa, appears to be immaterial. She functions as the Danish tourist's connection to all of sub-Saharan Africa.

The content analysis of the 17 tour operators websites reveals that 10 of these give absolutely no information about the local community in the respective destinations they serve. The aforementioned Nyati Safari describes the level to which locals are involved in tourism, indicating that '100 local staff [are] working to make sure that 50 guests can enjoy safari life at the luxury lodge.' The company gives financial support to the local school and also to appreciate the local residents it has created a funeral fund, which helps cover the expensive costs of the Shangan tribal funeral rituals. Nevertheless, Ving acknowledges that when an all-inclusive opens up, small local cafes and restaurants can have difficulties surviving. However, the company does not think this is a major issue because the all-inclusive resorts they work with tend to be isolated and no cafes and restaurants exist in the immediate vicinity. What is more visible, at least on the websites of some of the larger players are tabs linking to issues on social responsibility and the question as to whether or not all-inclusive travel is ethically correct. For instance, the website for Apollo talks about the importance for the travellers to reflect on how they travel and how they spend their money. They point out they use local commodities and that local labourers benefit.

Discussion

The title of this paper includes the question: does geography matter? We could add: does geography matter, to whom and in what way(s)? Though the all-inclusives present several elements, which give them the characteristics of being placeless, simultaneously they project feelings of safety, predictability, and comfort, 'a home away from home.' To be sure, tour operators have, on the one hand, the flexibility to easily substitute one all-inclusive with another as long as the latter offers a similar range of attractions and services. Yet, on the other hand, these players actually recognize the qualities of a variety of places, which at the end of the day causes them to employ a dynamic multi-locational strategy. Meanwhile, to locals the all-inclusive enclavic space certainly constitutes a (sociocultural) spatial entity that sharply differs from its surrounding environment. Clearly then, the answer to our question to a high degree depends on the way we contextualize the all-inclusive (focusing on the all-inclusive *per se* or placing it within a broader economic and geographical context) as well as our ability to shift perspective (perceiving the all-inclusive from the point of view of tourists, tour operators, or locals).

Destinations, especially those in regions of the world depending heavily on tourism, constantly pursue capital in the guise both of foreign direct investments but also the actions of tour operators and tourists themselves. Our study confirms that even if the destination or the accommodation facilities in an all-inclusive holiday package is spatially fixed, tour operators (representing international capital) are significantly mobile and flexible in the manner in which they organize their business. Tour operators adapt multi-locational as well as multi-product strategies, thereby minimizing their risks by not putting all their eggs into one basket. To attract the tour operators, the host country/destination must guarantee that the basic infrastructural needs for tourism (flight connections together with a certain standard regarding water supply, electricity, health care, etc.) can be satisfied. Even if it certainly takes more than physical infrastructure to stay competitive in a global and volatile market like tourism, this also ties into what David Harvey (1985) terms 'fixed capital,' that is capital transformed into a built environment, infrastructure, offices, houses, and other means to ensure production occurs.

However, since both tangible factors of production such as the built environment but also intangible factors of production like human capital are developed to suit a certain mode of production, undoubtedly these factors will eventually become obsolete. As a response to changing political or economic conditions, and in order to remain competitive, firms might seek out a more profitable location. For instance, a tour operator might abandon a resort that is not profitable enough and replace it with one that is supposedly more attractive to capital. Hence, capital – in the shape of tourists and tour operators – is mobile, while the destination, the small and medium enterprises or the all-inclusive resort is spatially fixed. Also, the exogenous tour operators are controlling the network of component suppliers that make up the all-inclusive holiday packages. This puts them in a position where they can choose for instance the hotels that give most value for the money, hence spurring competition. At a local level this might lead to *at least* two different roads of development; either to an improved quality of the tourism product or setting off a race to the bottom. Even if the tour operators keep their suppliers busy and ensure that the latter will make a profit, the power-relation remains asymmetrical (c.f. Koutoulas, 2006; Lee et al., 2015).

Table 1. All-inclusive continuum: type of tourism enclave, purpose of enclavism, and spatial characteristics.

Type of tourism enclave	Purpose of enclavism	Spatial characteristics
Cruise ships (e.g. to nowhere and cruises to private/leased islands or beaches).	Secure, comfort and predictability, keep 'the Other' outside. One price/spending onboard to maximize revenues, control quality and guest satisfaction.	Footloose, hypermobile resorts, free from regulations that bind land-based sectors. Placelessness with insignificance of place, standardized and 'international style'.
All-inclusive resorts (e.g. Maldives, Dominican Republic, Turkey and others).	Affordable, secure, comfort and predictability, keep 'the Other' outside. One price within the resort to maximize revenues, control quality and guest satisfaction.	Spatially fixed, decontextualized space and separated from the surrounding, dependency on foreign-controlled tour operators. Placelessness with insignificance of place, standardized and 'international style'.
All-inclusive holiday packages (e.g. Spain, Greece and others).	Affordable, convenience and predictability. One price to maximize revenues, control quality and guest satisfaction.	Spatially fixed, rarely isolated resorts. Destination and component suppliers dependent on foreign-controlled tour operators.

In Table 1, we sum up some of the traits of the different types of all-inclusives under study in this paper. The table is preferably read as a continuum of different types of all-inclusives and has been developed from an initial analysis of (1) type of tourism enclave; (2) purpose of enclavism; and (3) spatial characteristics.

Following this table overview, we argue that different kinds of tourism enclaves, in various ways, both control and are subjected to global interconnectedness, spaces of flows, and spatial fixity. At one end of the spectrum we have the *cruise ships to nowhere* whereas at the opposite end we can situate the *all-inclusive holiday package*. In between, lies the *all-inclusive land-based tourism enclave*. The *cruise ships to nowhere* represent the hypermobile enclosed floating resorts and a type of all-inclusive that also presents a significant example of who controls the flows and who/what is subjugated by the flows. Literally free-floating and rejecting spatial fixity, the shipping companies have found a way to 'cruise around' many of the laws and regulations stipulated by the nation-state (see also Wood, 2000). After all, while the cruise ship is *in flow*, it simultaneously holds a position from which it can exercise power over, for instance, flows of capital and labour. Perhaps this is what makes cruise ships to nowhere the chief symbol of elitist hypermobility.

Further to presenting this continuum, we also wish to highlight the issue of the importance of geographical location when it comes to all-inclusive enclavic models of tourism development. We have striven to indicate how all-inclusive holidays are marketed and through this identify how important or not place-based attributes are for the holiday product on sale. In the marketing of all-inclusive resorts, our study highlights the high level of disregard, especially among large-scale tour operators, for local spatial contingencies while services and facilities on offer are prioritized. As long as the all-inclusive offers free meals, outstanding pool areas and other recreational facilities, not to mention ample opportunities for activities and sports it matters very little where it is located. In other words, in such resorts there is an overriding tendency to downplay the destination's place-based attributes in pursuit of conveying to the visitors the image of attaining high-quality experiences in world-class environs.

In this regard, the all-inclusive resort emerges as a placeless phenomenon, and a decontextualized space. Hence, Relph's (1976) notion of placelessness serves as the basis for understanding homogenization and insignificance of place; places with loss of, or lack of, place identity. Placelessness refers to an illusory, enclosed, and decontextualized space that loses meaning and that diminishes the differences among places, such as for example tourist resorts. The development of 'vacationland,' 'consumer land,' and 'fantasy lands' (Relph, 1976, p. 99) are examples of utopias proving ideals to be copied, and as highlighted earlier in this paper, a direct counterpart to the all-inclusive resorts of activities and luxury standard. However, placelessness does not only include the physical environment itself, it also includes the underlying attitude that fails to acknowledge the significance of places. In that way, placelessness, in the meaning of anonymous spaces, homogeneous spaces, and interchangeable places, is the adoption of attitudes and the acceptance of mass values. Referring to Relph, Entrikin (1991) associates placelessness: (a) with modern society, its high-speed transportation and communication that makes the phenomenon of locality less significant, but also with; (b) the production of standardized landscapes and architecture of 'international-style' followed by the destruction of the meaning of places.

Concluding remarks

In this paper, we addressed the (non)importance of place and place-based attributes in contemporary mass tourism and explored how significant the actual geographic location (and surroundings) of an all-inclusive holiday is and whether this is overshadowed by other factors such as accommodation facilities, activities on offer, or value for money among those who market all-inclusive holidays. We conclude that contrary to certain contemporary tourism trends, which emphasize the full immersion of tourists within the host society, an all-inclusive holiday resort, through its scale, design and function, purposefully isolates itself from its immediate surroundings. In such a property, there appears to be a genuine lack of interest by its operators as well as those who market it to encourage visitors to venture outside the gates and mix with the host society. A common feature of many all-inclusives worldwide is their enclavic design, the very fact that these are enclosed and secure spaces. As the Scandinavian tour operators in this study claim, this infuses a sense of safety and comfort among the tourists. The all-inclusive, at its extreme, works as a means of keeping 'the Other' at bay. Therefore, the all-inclusives are spaces designed specifically for tourism and target outsiders, while simultaneously, these enclosed spaces become all-exclusive zones for local communities. To be sure, in some instances, this barrier is somewhat porous, since workers from the outside are brought to the inside but to a major extent their presence is temporary and involves minimum interaction with the guests.

We conclude that geography does not matter, or matters minimally, in the production and marketing of all-inclusives and the experiences they entail. It is placeless representations of space and not specific geographic characteristics that are highlighted to sell travel to the all-inclusive enclavic sites. These representations are produced and distributed largely from the outside, thus having minimal bearing on how local people understand and use their own land and resources. Rather the pictures and symbols used to market all-inclusive enclaves reflect a cosmopolitan and affluent way of life. The all-inclusives pay very little attention to the importance of place-based attributes and identities and as such they represent a decontextualized and interchangeable space. However, this does not automatically mean that they are not meaningful to those who visit them. After all, these enclaves can be sites for valued family reunions or ones were couples get married and spend their honeymoons or anniversaries and, thus, in their own right they help generate holiday memories of positive experiences of the all-inclusive holiday.

The study's findings derive primarily from the manner in which a key group of gatekeepers (i.e. tour operators) manipulate flows to enclavic spaces. However, the study also proves that more can be done on this topic. For instance, future research should investigate the types of linkages and flows permeating the walls and borders of the enclaves. This would enable deeper understanding of the manner in which enclavic spaces interface with their surroundings. For example, researchers could investigate tourists' perceptions of what lies beyond the enclave and whether or not there are place-based attributes that would indeed encourage them to escape the enclave's perimeter. Meanwhile, examinations of linkages that do exist between the enclave and its surroundings in the form of flows of workers, goods, and capital must also be undertaken. We believe that more focus should be placed on studies of the implications that resort enclaves have for local host societies since these, in turn would lead to a superior understanding of how to develop and implement effective destination management strategies. In relation to this, critical research on policy, planning, and governance aspects of enclave developments is needed.

In final analysis, we believe that tourism researchers must expand their focus more on the divergent trends affecting contemporary tourism, paying particular attention toward whether or not travelers' motives to participate in an inclusive holiday at a resort enclave are motivated by place-bound geographic characteristics or mostly by the amenities and activities on offer. Do these visitors view the enclave as a destination in its own right regardless of its surroundings or are they intrigued by its geographical surroundings? Are there variations in the types of visitors who visit these enclaves based on their motivations? More knowledge about factors in processes of change, including tourism consumption and travel behaviour, are important for the understanding of transformations and (re) shaping of destinations.

Acknowledgements

We would like to thank Sanna Fors and Mikael Wassdahl for assisting in data collection. We are also grateful to the anonymous reviewers for their constructive suggestions.

Disclosure statement

No potential conflict of interest was reported by the authors.

References

Aili, L., Jiaming, L., & Min, L. (2007). Progress in enclave tourism study of overseas: A literature review. *Chinese Journal of Population Resources and Environment, 5*(3), 76–81.

Andersen, W. (2011). Enclave tourism and its socio-economic impact in emerging destinations. *Anatolia: An International Journal of Tourism and Hospitality Research, 22*(3), 361–377.

Arefi, M. (2007). Non-places and placelessness as narratives of loss: Rethinking the notion of place. *Journal of Urban Design, 4*(2), 179–193.

Bagaeen, S. (2007). Brand Dubai: The instant city, or the instantly recognizable city. *Internationall Planning StudiesInternational Planning Studies, 12*(2), 173–197.

Birkeland, I. (2008). Cultural sustainability: Industrialism, placelessness and the re-animation of place. *Ethics, Place & Environment, 11*(3), 283–297. https://doi.org/10.1080/13668790802559692.

Bladh, C., & Holm, H. (2013). Can economics explain where all-inclusive deals are offered? *Tourism Economics, 19*(2), 339–348.

Boyle, D. (2016, May 17). World's largest cruise ship, Harmony of the Seas, arrives in Southampton. *The Telegraph.* Retrieved from http://www.telegraph.co.uk/news/2016/05/17/worlds-largest-cruise-ship-harmony-of-the-seas-arrives-in-southa/.

Britton, R. (1979). The image of the Third World in tourism marketing. *Annals of Tourism Research, 6* (3), 318–329.

Britton, S. G. (1982). The political economy of tourism in the Third World. *Annals of Tourism Research, 9*(3), 331–358.

Bunnell, T., Muzaini, H., & Sidaway, J. D. (2006). Global city frontiers: Singapore's Hinterland and the contested socio-political geographies of Bintan, Indonesia. *International Journal of Urban and Regional Research, 30*(1), 3–22.

Carnival. (2014). Retrieved from http://www.carnival.com/cms/fun/destination/Nowhere/cruise_to_nowhere.aspx?subRegionCode=CN

Cohen, E. (1972). Rethinking the sociology of tourism. *Annals of Tourism Research, 6*(1), 18–35.

Conrad, S. (2011). What does all inclusive mean? All Inclusive Outlet Website. Retrieved from http://www.allinclusiveoutlet.com/news/travel-tips-and-vacation-planning/what-does-all-inclusive-mean/

Cresswell, T. (2004). *Place: A short introduction*. Malden: Blackwell.

Dielemans, J. (2008). *Välkommen till Paradiset: Reportage om turistindustrin*. Stockholm: Bokförlaget Atlas.

Domroes, M. (2001). Tourism in the Maldives: The advantages of the Resort Island concept. *Tourism, 49*(4), 369–382.

Douglass, M., & Huang, L. (2007). Globalising the city in Southeast Asia: Utopia on the urban edge – the case of Phu My Hung, Saigon. *International Journal of Asia Pacific Studies, 3*(2), 1–42.

Edensor, T. (1998). *Tourists at the Taj. Performance and meaning at a symbolic site*. London and New York, NY: Routledge.

Edensor, T. (2007). Mundane mobilities, performances and spaces of turism. *Social & Cultural Geography, 8*(2), 199–215.

Entrikin, J. N. (1991). *The betweenness of place. Towards a geography of modernity.* Baltimore, MD: Johns Hopkins University Press.

Foglesong, R. (2001). *Married to the Mouse: Walt Disney World and Orlando*. New Haven: Yale University Press.

Freestone, R., & Liu, E. (2016). *Place and placelessness revisited*. New York, NY: Routledge.
Freitag, T. G. (1994). Enclave tourism development: For whom the benefits roll? *Annals of Tourism Research, 21*(3), 538–554.
Gui, L., & Russo, A. P. (2011). Cruise ports: A strategic nexus between regions and global lines – Evidence from the Mediterranean. *Maritime Policy & Management, 38*(2), 129–150.
Hallett, R., & Kaplan-Weinger, J. (2010). *Official tourism websites: A discourse analysis perspective*. Bristol: Channel View Publications.
Harvey, D. (1985). *The urbanization of capital*. Baltimore, MD: John Hopkins University Press.
Harvey, D. (2013). *Rebel cities: From the right to the city to the urban revolution*. London: Verso.
Ioannides, D. (1998). Tour operators: The gatekeepers of tourism. In D. Ioannides & K. Debbage (Eds.), *The economic geography of the tourism industry: A supply-side analysis* (pp. 139–158). London: Routledge.
Ioannides, D., & Debbage, K. (1998). *The economic geography of the tourism industry: A supply-side analysis*. London: Routledge.
Issa, J. J., & Jayawardena, C. (2003). The 'all-inclusive' concept in the Caribbean. *International Journal of Contemporary Hospitality Management, 15*(3), 167–171.
Jaakson, R. (2004). Beyond the tourist bubble? Cruise ship passengers in port. *Annals of Tourism Research, 31*(1), 44–60.
Judd, D. (1999). Constructing the tourist bubble. In D. Judd & S. Fainstein (Eds.), *The tourist city* (pp. 35–53). New Haven: Yale University Press.
Koutoulas, D. (2006). The market influence of tour operators on the hospitality industry. In A. Papatheodorou (Ed.), *Corporate rivalry and market power* (pp. 94–123). London: I.B Tauris.
Law, R., & Chen, S. Z. (2012). Representation of destination cultural factors on hotel websites: Content analysis of Beijing hotel websites. *Asia Pacific Journal of Tourism Research, 17*(2), 210–229.
Lee, D., Hampton, M., & Jeyacheya, J. (2015). The political economy of precarious work in the tourism industry in small island developing states. *Review of International Political Economy, 22*(1), 194–223.
MacLeod, N. E. (2006). The placeless festival: Identity and place in the post-modern festival. In D. Picard & M. Robinson (Eds.), *Festivals, tourism and social change: Remaking worlds* (pp. 222–237). Clevedon: Channel View Publications.
Manuel-Navarrete, D. (2016). Boundary-work and sustainability in tourism enclaves. *Journal of Sustainable Tourism, 24*(4), 507–526.
Massey, D. (1991). A global sense of place. *Marxism Today, 39*, 24–29.
Massey, D. (2005). *For space*. London: Sage Publications.
Mattsson, K. (2010). Semesterutopier. In M. Tesfahuney & K. Schough (Eds.), *Det globala reseprivilegiet* (pp. 79–91). Lund: Sekel.
Mbaiwa, J. E. (2005). Enclave tourism and its socio-economic impacts in the Okavango Delta, Botswana. *Tourism Management, 26*, 157–172.
Meyer-Arendt, K. J. (1992). Seaside resorts in the Dominican Republic: A typology. *Journal of Geography, 91*, 219–225.
Mosedale, J. (2016). Neoliberalism and the political economy of tourism: Projects, discourses and practices. In J. Mosedale (Ed.), *Neoliberalism and the political economy of tourism* (pp. 1–20). Oxon and New York, NY: Routledge.
Myers, G. N., & Jainchill, J. (2015, October 7). The all-inclusive evolution. *Travel Weekly*, pp. 1–6.
Relph, E. (1976). *Place and placelessness*. London: Pion.
Reynolds, J. (2015). *11 new (and newly renovated) all-inclusive resorts that need to make your 2016 visit list*. Retrieved from https://www.oyster.com/articles/50290-11-new-and-newly-renovated-all-inclusive-resorts-that-need-to-make-your-2016-visit-list/
Rodrigue, J.-P., & Notteboom, T. (2013). The geography of cruises: Itineraries, not destinations. *Applied Geography, 38*, 31–42.
Rowley, J., & Slack, F. (1999). The retail experience in airport departure lounges: Reaching for timelessness and placelessness. *International Marketing Review, 16*, 363–375.

Royal Caribbean International. (n.d.). *Labadee®: The ultimate provate beach: Only Royal Caribbean can take you there*. Retrieved from http://www.royalcaribbean.com/findacruise/ports/group/home.do?portCode=LAB
Saarinen, J. (2017). Enclavic tourism spaces: Territorialization and bordering in tourism destination development and planning. *Tourism Geographies, 19*(3), 425–437.
Schyst Resande. (2011). Utsugning av vissa, guldkant för andra? En studie av all-inclusive turism och den svenska turistbranschen. Retreived fromwww.schystresande.se
Schyst Resande. (2012). *Glitter och glamor – turismen i resejournalistiken*. Retreived fromwww.schystre sande.se
Shaw, B. J., & Shaw, G. (1999). 'Sun, sand and sales': Enclave tourism and local entrepreneurship in Indonesia. *Current Issues in Tourism, 2*(1), 68–81.
Shim, C., & Santos, C. A. (2014). Tourism, place and placelessness in the phenomenological experience of shopping malls in Seoul. *Tourism Management, 45*, 106–114.
Sidaway, J. (2007). Enclavic space: A new metageography of development? *Area, 39*, 331–339.
Silverman, D. (2013). *Doing qualitative research: A practical handbook*. London: Sage.
Swedish Chambers of Commerce. (2011). *Market report: Focus on the EU and Swedish Market: Outbound tourism*. Stockholm: Swedish Chambers of Commerce.
TNS Political & Social. (2015). *Preferences of Europeans towards tourism. Flash Eurobarometer*. European Commission, Directorate-General for Internal Market, Industry, Entrepreneurship and SMEs. European Union. Retreived from http://ec.europa.eu/commfrontoffice/publicopinion/flash/fl_414_en.pdf
Van Wijk, J., & Persoon, W. (2006). A long-haul destination: Sustainability reporting among tours operators. *European Management Journal, 24*(6), 177–189.
Warren, S. (1999). Cultural contestation at Disneyland Paris. In D. Crouch (Ed.), *Leisure/tourism geographies: Practices and geographical knowledge* (pp. 109–125). London: Routledge.
Weaver, A. (2005). Spaces of containment and revenue capture: 'Super-sized' cruise ships as mobile tourism enclaves. *Tourism Geographies, 7*(2), 165–184.
Wong, C. S., & Kwong, W. Y. (2004). Outbound tourists' criteria for choosing all-inclusive package tours. *Tourism Management, 25*, 581–592.
Wood, R. (2000). Caribbean cruise tourism: Globalization at Sea. *Annals of Tourism Research, 27*(2), 345–370.

Websites of Scandinavian tour operators included in the study for content analyis:

http://www.adventureholidays.dk
http://www.afrikasafari.no/
http://www.apollo.se
http://www.detur.se/
http://www.enjoycaribbean.com/ (not accessible in 2016)
http://www.fritidsresor.se
http://www.hummingbirdtravel.se/
http://www.indiatravels.se/ (not accessible in 2016)
http://www.kenzantours.se/
http://www.nazar.se/
http://www.nyati-safari.se/
http://www.qualitours.se/
http://www.solia.no/
http://www.solresor.se/
http://www.turkietresor.se/
http://www.usaspesialisten.no/
http://www.ving.se

Interviews

Apollo, Communication manager, 27 April 2011
Apollo, PR & Social Media Manager, 27 October 2016
Detur, senior sales executive, 17 August 2011
Detur, Sales & Marketing Coordinator, November 2016
Fritidsresor/TUI, CSR coordinator, 2 May 2011
India Travel, Vice director, 12 May 2011
Ving, Communication manager, 5 May 2011
Ving, Communication manager, 18 October 2016

Selling bubbles at sea: pleasurable enclosure or unwanted confinement?

Adam Weaver

ABSTRACT
The insularity of super-sized cruise ships is simultaneously part of their appeal and a problem. Seagoing tourism enclaves offer a world of fantasy, consumption, familiarity, and diversion. However, problems encountered by passengers with respect to seasickness, boredom, and a sense of entrapment are noted within a range of publications. Selling the desirable aspects of cruise-ship enclosure co-exists with attempts to address the unease, discomfort, and uncertainty associated with bounded domains. These two currents, to use a nautical metaphor, characterize – or steer – the way in which shipboard experiences are sold to passengers. The concept of place marketing can be related to the thrust of the currents. Place marketing has been widely addressed in relation to urban and rural areas but not tourism enclaves such as cruise ships. The insular cruise-ship environment – a contained, consumption-driven place – is sold to (prospective) passengers in different ways. Those who study place marketing have yet to address dissimilar, but complementary, promotional messages (or currents) that are used to sell resort-style environments. Commercial imperatives explain the actions of cruise-ship companies with respect to selling the enclosed nature of their vessels, especially to North American consumers. The way in which place marketers manage place-based images – a response to intensified intercity or inter-town competition – mirrors the careful management of the promotion of shipboard spaces, often a response to intense competition within the cruise industry and part of initiatives designed to attract first-time passengers. This study addresses a research question: How are different but congruous promotional messages deployed as part of efforts to sell shipboard spaces?

Introduction

A distinct aspect of the present-day cruise industry is the recent growth in the size and number of its vessels. Many of the world's most prominent cruise-ship brands – Carnival Cruise Lines, Royal Caribbean International, Princess Cruises, Norwegian Cruise Line, and Disney Cruise Line – operate ships that can accommodate over 3000 passengers (Coffman, 2014; Ward, 2017). These ships are a manifestation of the Las Vegas and Disney logic of offering a total, fully immersive experience of excitement, entertainment, and shopping. Promotional endeavors typically emphasize that the ships – especially the super-sized vessels (Weaver, 2005a) – are places of comfortable familiarity and plenty. The pleasure-filled, idealized world of the cruise-ship bubble can also, for some, offer excessive confinement. Selling the sought-after features of enclosure takes place alongside selling-related endeavors that address the more problematic aspects of the bubble. The cruise industry has to contend with varying attitudes towards its vessels – that those on board may find that they are personally incompatible with the co-occupants of the vessel or believe that they will be bored, become seasick, or contract a virus during their holiday. Tourist bubbles can therefore be a source of worry for those considering a cruise vacation. In response to these concerns, mass-market and trade publications that support the selling of cruise vacations convey messages of reassurance regarding the nature of seagoing bubbles (Cullinan, 2016; Eisen, 2007a, 2008; Young, 2015). That there are 'troubles with bubbles' is recognized; certain types of spaces – under certain circumstances – are not necessarily perceived favorably. This study represents an extension of existing cruise-related marketing research by examining the promotion of contained shipboard space rather than the more frequently explored theme of cruise-ship passengers' characteristics – for example, their social media use (Park, Ok, & Chae, 2016), the attributes of first-time and repeat passengers (Petrick, 2004), and behavioral intentions (Duman & Mattila, 2005). For cruise-ship companies, there is a need to sell the advantageous attributes of shipboard space while also challenging negative perceptions of containment. Research that addresses the way in which particular types of tourism-oriented spaces, such as super-sized cruise ships, are sold to consumers should be of interest to scholars curious about intersections between geography and commerce.

The study of place marketing has typically addressed attempts to make cities, towns, and more remote destinations attractive sites for external investment and tourism (Coles, 2003; Hall, 2008; Schöllmann, Perkins, & Moore, 2001; Smith, 2005). Place marketing has the potential to reinvent or re-image cities and towns (Smith, 2005); places are carefully packaged for tourism consumption. A similar phenomenon is applicable to super-sized cruise ships; the insularity of the cruise ship is promoted in different ways. Sometimes the desirable features of enclosure are emphasized. In other instances, cruise-ship companies project messages that are designed to address concerns about shipboard containment. That first-time passengers, for example, perceive cruise ships as too confining potentially undermines the strategic ambitions of cruise-ship companies seeking to broaden the appeal of their product (Cullinan, 2016; Eisen, 2007a, 2008; Young, 2015). The drive to earn profit prompts cruise-ship companies to adapt their marketing messages – even crafting different messages that represent different currents – in order to sell enclosure. At the core of the study is a research question that combines marketing- and enclave-related considerations: How are different, but still compatible, promotional messages used to sell the circumscribed spaces of cruise-ship bubbles? This study acts as an invitation for more research that explores the selling of tourism-oriented spaces, including enclave environments.

A crucial relationship underpins the selling of bubble environments at sea. This relationship can be compared to two currents that flow slightly differently but ultimately move in the same destination. They are complementary in terms of the commercial purpose they serve. One current pushes in a direction that stresses the desirable features of the cruise-ship bubble; the other has a different orientation and offers reassurance that aims to counteract unfavorable perceptions of bubbles. Far from being diametrically opposed to each other, the two currents – despite their differences – function compatibly with each other and have the same overarching objective: selling cruise vacations. The cruise-ship bubble is situated at the confluence of different currents that move differently but have a similar, business-driven imperative.

This introduction is followed by a review of recent works that address several relevant themes – namely, the practice of place marketing, the nature of tourist bubbles, the connection between consumption and unease, and the way in which consumption-related activity reflects certain complex connections between interwoven phenomena. Sources of data are then described as well as the method of analysis used. The data were obtained from cruise-vacation guidebooks as well as articles obtained from two sources: a magazine that targets cruise enthusiasts, *Cruise Travel*, and various travel-oriented trade publications. These sources identify the means by which bubble environments are sold to prospective passengers. The findings of the study are subsequently documented and organized around the idea that two broad themes – described as currents – inform the selling of sea-going bubble environments. One current depicts these bubble environments as bountiful, but there is also 'bubble trouble'; the marketing of cruise-ship space must address concerns related to a sense of excessive enclosure. A final section offers conclusions and directions for future scholarly enquiry.

Literature review

Tourist bubbles are spaces within which the needs and preferences of tourists, almost exclusively, are served. Within these bubbles, tourists can experience comforting familiarity while abroad during their holiday (Cohen & Neal, 2012; Gutberlet, 2016; Jaakson, 2004; Minca, 2000; Saarinen, 2017). Enclave resorts and theme parks, as well as cruise ships, offer an immersive experience explicitly designed for consumption (Jacobsen, 2003; Kolberg, 2013; Lukas, 2008). They remove visitors from their day-to-day existence, providing them with access to a world of pleasurable diversions. Recent work has documented the emergence and proliferation of super-sized cruise ships, bounded seagoing bubbles that inspire a degree of awe due to their size and expanding array of facilities (Vogel & Oschmann, 2013; Weaver, 2005a; Wood, 2000). These vessels offer structured, standardized hospitality and entertainment (Weaver, 2005b). Shipboard spaces may be sites of inter-passenger social interaction (Yarnal, 2004; Yarnal & Kerstetter, 2005) but they are also domains where business-directed practices and processes operate.

Selling space – or, more precisely, selling enclosure – can be tied to research that addresses place marketing. Place marketing reflects an entrepreneurial ethos whereby an array of urban and rural environments are promoted through the use of cultural districts, prestige 'flagship' projects, and other noteworthy attributes that convey a desirable image (Coles, 2003; Hall, 2008; Schöllmann et al., 2001; Smith, 2005). The marketing of various types of places has received a fair amount of attention from researchers, with considerable criticism of the selective ways in which cities are promoted (Kearns & Philo, 1993; Ward, 1998). Certain place-based images are promoted in favor of others. The marketing of cities, towns, and other domains mirrors the marketing of corporate-owned tourism enclaves such as cruise ships. Tourism-oriented spaces that are bounded domains – and do not have a (potentially vocal) permanent population – are also adroitly packaged and sold. The promotion of shipboard space within the cruise industry addresses a dilemma with respect to insularity. A confined environment provides comfort through the provision of familiar amenities but it is also a source of concern because it has the perceived potential to create a sense of isolation and entrapment. The selling of cruise-ship bubbles stresses the pleasures of containment but it must also address perceived problems related to confinement. To date, cruise-related marketing research has mostly addressed the attributes and actions of passengers rather than the strategies used to sell contained shipboard spaces (Duman & Mattila, 2005; Larsen, Wolff, Marnburg, & Øgaard, 2013; Park et al., 2016; Petrick, 2004). Instead of exploring demand-related matters and the experiences of passengers, the selling of space is the focus of attention.

Enclosed enclaves that have a conspicuous presence within the marketplace are not free of controversy. Tourist bubbles have been subjected to criticism because the revenue they capture rarely produces noticeable benefits for nearby communities (Freitag, 1994). However, some commentators note that restricting tourists' activities to the premises of the enclave may protect the natural environment, minimize so-called cultural contamination, and create a vibrant node of entrepreneurial opportunities for nearby residents (Jaakson, 2004; Nunkoo & Ramkissoon, 2016). Favorable and unfavorable perceptions of the tourist bubble not only exist within scholarly studies but are also visible within publications that promote cruise vacations. Exploring the promotion of cruise-ship bubbles exposes a form of unease tied to consumption, specifically in relation to insularity and spatial containment.

Bubble-based consumption can be a cause for concern. It is a source of pleasure and problems. The relationship between consumption and the doubts and worry it creates has been studied extensively (de Botton, 2004; Kirillova, Lehto, & Cai, 2017; Warde, 2004). There is unease tied up in an activity meant to be a source of amusement and gratification. Some research situated at the consumption-worry nexus coalesces around food (Jackson, 2010). Other authors have observed that pervasive consumerism is responsible for a great deal of concern amongst academic commentators (Horowitz, 2004, 2012). A host of worrisome moral and cultural issues have arisen regarding mass consumption, ranging from the widespread acceptance of free market capitalism and the ubiquitous nature of advertising to whether intensive materialism promotes excessive hedonism and conceited status seeking. Many problems associated with consumption are also associated with the hazards they pose to the natural environment (Horowitz, 2004). The concerns associated with cruise-ship bubbles, especially within publications where the selling of such spaces rises to the surface, relate to the bounded nature of these vessels when at sea rather than to a deeper consideration of the broader, societal repercussions of shipboard super-abundance. However, worries associated with confinement, as mundane as they may seem, should not be dismissed. A ship at sea is in a vulnerable position should an accident occur, a phenomenon demonstrated by the sinking of the *Costa Concordia* in early 2012 (Volo & Pardew, 2013).

Studies of consumption expose an array of complex interrelationships (Su, 2010). These relationships include those that exist between freedom and control as well as those underpinning the desire to maintain a distinctive personal identity and still be part of a wider, collective community. Although the consumption of tourism involves experiences related to freedom and creativity, the provision of these experiences through the medium of the market undermines the realization of freedom and the potential for human agency (Miles, 2010; Wood, 1994). Freedom and control are therefore deeply interrelated. Similarly, the notion that identity construction is an individual enterprise is entangled with the role performed by peer appraisal and the desire to conform (Yan & Santos, 2009). Self-expression through consumer choice occurs within the context of societal perceptions and norms. The cruise-ship bubble, a particular type of consumption-driven environment, is the focal point of a complex relationship – one between complementary and compatible currents – whereby the vessel at sea has the potential to provide both pleasurable encapsulation and worry- or uncertainty-creating confinement.

Methods

Published sources such as guidebooks, a cruise-enthusiast magazine, and travel-related trade journals are used to explore the selling of cruise-ship bubbles. The rise of mass tourism was made possible by advances in transportation technology, the development of commercial and communication networks, and the emergence of enterprises that bundled together various services and created a sophisticated logistics of travel. Travel guidebooks and magazines are part of this machinery underpinning the expansion of mass tourism (Iaquinto, 2011; Peel, Sørensen, & Steen, 2012); they connect the needs of the tourism industry with readers' desires to best exploit their holiday time. As travel guidebooks and magazines occupy a domain comparable to Goffman's (1959) front regions – plainly visible to consumers – they present widely circulated images of tourism. These front-region sources, in the context of cruise travel, are easily accessible to both first-time and repeat passengers.

Travel-oriented trade publications feature content that is primarily of interest to industry observers and practitioners (Carpenter & Upchurch, 2008; Coomber, Lee, Lee, & Cairncross, 2006), a Goffman-esque back region of commercial information. For industry professionals, trade journal articles offer advice and provide information outside the scope of personal knowledge. The issues that these articles address are mainly not of interest to casual readers or even cruise passengers. Despite their different audiences, front- and back-region sources address the selling of cruise vacations and, more specifically, shipboard bubble environments.

The cruise-vacation guidebooks consulted for this paper were produced by Berlitz and Fodor's, two companies renowned for their travel-related publications (Coffman, 2014; Ward, 2017). Fifteen articles from a magazine produced for cruise enthusiasts, *Cruise Travel*, were also examined. Past issues of this magazine during a 10-year period, between 2007 and 2016, have featured articles describing a series of super-sized cruise ships. The articles were accessed via *Hospitality and Tourism Complete*, a searchable database that contains both scholarly and industry-focused content. A mass circulation magazine such as *Cruise Travel*, similar to widely available guidebooks, helps to foster the impulse to cruise by projecting accessible, front-region depictions of this type of vacation. Travel guidebooks and magazines both promote cruise travel and offer expert or editorial advice to (prospective) passengers. The trade journal articles, similar to the articles published in *Cruise Travel*, were obtained from *Hospitality and Tourism Complete*. Twenty-two articles from travel-oriented trade publications such as *Travel Agent, Travel Weekly*, and *Travel Trade Gazette UK & Ireland* were analyzed during a 10-year period (2007 to 2016) (Table 1).

The analysis of the various written texts was undertaken with a sensitivity towards the different ways in which cruise-ship bubbles are depicted. There have been careful appraisals of written texts that have a connection with cruise tourism. Passengers' experiences with shipboard spaces have been examined through a study of Internet-based discussion board content (Kwortnik, 2008). In a manner similar to the way in which discussion board comments were read carefully by Kwortnik (2008), published sources were reviewed thoroughly for the present research. The repeated reading of the published sources demonstrates that the selling of cruise-ship bubbles is a complex endeavor that takes into account the merits and drawbacks of enclosure. Slightly different and yet compatible approaches to selling these bubbles would appear to be co-present. The idea that the selling of cruise vacations is shaped by two metaphorical currents that move, in broad terms, in the same direction was introduced as a means of organizing the findings conceptually. Both front- and back-region publications discuss the comfort and familiarity offered by enclaves as well as problems and concerns related to shipboard confinement. The small sample means that applying techniques associated with quantitative analysis would not be appropriate. Compared to a method that merely counts the occurrence of certain words, the type of analysis conducted in this study enables a deeper study of meanings (Knudsen, Soper, & Metro-Roland, 2007).

One crucial strength of this text-based study resides in the detailed descriptions of cruise-ship environments within the sources consulted. These sources, whether targeting

Table 1. Inventory of publications analyzed.

Publication	Number of items analyzed
• Cruise vacation guidebooks	• Two guidebooks, one published by Berlitz and the other by Fodor's
• *Cruise Tourism* (a cruise-focused travel magazine)	• Fifteen articles
• *Travel Agent* (a trade publication)	• Seven articles
• *Travel Weekly* (a trade publication)	• Seven articles
• *Travel Trade Gazette UK & Ireland* (a trade publication)	• Six articles
• *Canadian Travel Press* (a trade publication	• One article
• *Home-Based Travel Agent* (a trade publication)	• One article

consumers or industry practitioners, offer a means to understand widely projected images of cruise vacations and the manner in which they are sold. However, a text-centric approach to research privileges the academic analyst's reading of the material while ignoring the perspectives of intended readers. Scholarly interpretations of texts have the potential to differ from the meanings extracted by non-academic readers (Marchand, 1985; O'Barr, 1994). Reading travel guidebooks and magazines such as *Cruise Travel* for the purpose of making a purchase – or reading trade journal articles in order to extract business- or industry-related intelligence – is not the same as reading those same texts in order to understand the selling of shipboard spaces.

There are a number of alternative ways to read textual material, and a number of possible subject-positions from which to understand it (Knudsen et al., 2007). A critical, academic reading of consumer-focused and industry-oriented texts strives to expose meanings that are not always immediately obvious. Text-based content, and its complexities, are combined with a clear analytical voice. This voice, through the use of a nautical metaphor, provides expression to both the more desirable aspects of insularity as well as attempts by cruise-ship companies to counteract concerns connected with containment.

Findings

Bountiful bubbles

The cruise industry offers passengers comfort and enclosure. Cruise-ship bubbles are climate-controlled, revenue-generating domains, enormous in size and offering an array of services and amenities. The artificiality of the cruise-ship environment and the sense of separation from the outside world – exemplified by Disneyfied themed settings on-board as well as a mainstreamed Las Vegas ostentatiousness – is complemented by attempts to create pleasant familiarity. In other words, the analyzed texts demonstrate that shipboard spaces carefully straddle a boundary between being sheltered, inward-facing realms and embracing certain aspects of external, off-ship environments. The marketing current that embraces and emphasizes the benefits of enclosure references the outside world and, at the same time, tries to shut it out.

Selling cruise vacations sometimes involves making connections or comparisons with other enclosed realms, typically ones with which most passengers would be well acquainted – for instance, shopping malls. A number of ships operated by Royal Caribbean International (RCI) feature 'indoor mall-like environment[s]' (Ward, 2017, p. 429) and have been described as 'the floating equivalent of a shopping mall' (Ward, 2017, p. 218) by one author of a cruise-vacation guidebook. These ships, similar to shopping malls, are designed primarily to influence people to purchase products and services. The selling of cruise-ship bubbles, in addition to referencing familiar retail spaces, emphasizes a sense of totality. This marketing current sometimes stresses a desirable comprehensiveness. Prior to the maiden voyage of *Regal Princess*, the actor who portrayed Captain Merrill Stubing in the television series *The Love Boat*, Gavin MacLeod, made a remark published in the magazine *Cruise Travel*:

> I just can't believe what a beautiful, comfortable ship this is. And it's so huge! It's got everything you could possibly ever want – and then some. (Knego, 2015, pp. 29–30)

Efforts are made to surround passengers with a familiar plenitude that simultaneously replicates aspects of their day-to-day existence and aims to exceed their expectations. Shipboard spaces are portrayed, via this marketing current, as offering tremendous abundance.

The themes of familiarity and plenty are compatible with frequent references made to the urban environment in a range of cruise-related publications. *Allure of the Seas, Harmony of the Seas*, and *Oasis of the Seas* – ships operated by RCI – are structured into seven named 'neighborhoods': Central Park, the Boardwalk, the Royal Promenade, the Pool and Sports Zone, Vitality at Sea Spa/Fitness Center, Entertainment Place, and Youth Zone. Each of these neighborhoods has an identity and an associated array of features and activities. The neighborhoods are 'distinguished by purpose (spa and fitness, pool and sports), age, (youth zone), design (Central Park and the Boardwalk), or function (entertainment)' (Coffman, 2014, p. 286). Central Park is a terrestrial environment transplanted onto a waterborne cruise ship; it 'is 328 ft (100 m) long, and has 27 real trees and almost 12,000 plants, including a vertical "living" plant wall' (Ward, 2017, p. 561). This neighborhood-oriented naming strategy positions these ships as equivalent to cities. The contained environments of the ships are portraying as having the around-the-clock, full-scale vibrancy of a cosmopolitan, urban environment. Seagoing tourist bubbles are sometimes compared to shopping malls, and sometimes to cities. However, such comparisons consistently convey the message that cruise ships are consumption-driven environments that feature copious choice – in excess of that available at home – as well as experiences with which consumers are already familiar. This marketing current flows with a tendency to provide positive, heartening comparisons.

Bubbles at sea, similar to cities, have exclusive enclaves that provide greater privacy and enhanced opportunities to spend. There are bubbles within bubbles; the enclosure typically offered by cruise ships, as a whole, may not offer the degree of exclusivity some passengers desire. For passengers seeking an immersive experience that exceeds the type of exclusivity and privilege that is usually available on board, '"gated" areas are for those paying extra to live in a larger suite and gain access to "private" facilities, concierge lounges, and private sunbathing areas' (Ward, 2017, p. 23). Princess Cruises operates several ships that feature 'The Sanctuary' on board (Seldon & Seldon, 2013). Similar shipboard spaces referred to as 'The Haven' have been established by Norwegian Cruise Line (Leppert, 2016b). This bubble-within-a-bubble phenomenon could be interpreted as the function of efforts by the more mainstream cruise-ship companies to compete with companies that target higher spending passengers. The development of these shipboard enclaves could be considered a corporate response to an inclination expressed by some passengers to secure seclusion within busier, increasingly densely populated vessels. Some passengers seek comfort within restricted-access spaces that resemble the types of enclosed, controlled-entry domains where they may reside or work. The insularity that is the result of a ship's presence at sea can be supplemented with the purchase of additional insularity – a heavily-promoted, extra-fee amenity. This dimension of the 'bountiful bubble' marketing current takes the notion of enclosure further; even greater cloistering is available to those who seek it and can afford it.

Brands and branding are also emblematic of familiarity and plenty. Prominent brands connect the realm that exists outside of the cruise-ship bubble with the internal hyperreality of the ship. Tourist bubbles insulate passengers from unwanted external intrusions but they also offer opportunities for brands to have exclusive access to customers for an extended duration. Several brands have established a presence on board ships operated by Carnival Cruise Lines. The cruise-ship company has partnered with 'EA SPORTS [a maker of sports-themed video games] to create EA SPORTS Bars at sea and Guy's Burger Joint, [a venue developed] in partnership with [television personality and] *Food Network* star Guy Fieri' (Coffman, 2014, p. 154). One of the Disney-operated ships, *Disney Dream*, has an 'interactive play area' (Cruz, 2016, p. 14) infused with elements derived from the highly successful *Star Wars* film franchise. In-cabin television options on board many cruise ships include the readily recognizable broadcasting brands – for example, CNN and ESPN (Ward, 2017). The presence of Coca-Cola on board a number of ships has been augmented by the installation of beverage dispensing 'machines that offer 165 soft-drink flavors' (Leppert, 2016a, p. 20). Cruise-ship bubbles are branded bubbles. Containment at sea does not mean that sacrifices, such as in relation to one's favorite products or brands, have to be made. 'Bountiful bubbles' are designed to embrace those aspects of the outside world that are deemed to be desirable, and products emblazoned with familiar brand names are an important component of the marketing current that portrays cruise ships as enclaves of enticements.

The metaphorical current that has been explored so far – one that is consistent with the themes of familiarity, comfort, and abundance – emphasizes the desirable aspects of enclosure. However, a current that is simply associated with selling idealized attributes of the cruise-ship bubble does not account for a steady stream of enclosure-related problems that are pre-emptively addressed by various publications. This steady stream is connected to a second current that responds to passengers' concerns regarding the unruly co-occupants of the ship, boredom, the potential for seasickness, and communicable viruses. A current that attempts to wash away these problems may specifically target first-time passengers as they may have the most serious bubble-related uncertainties about cruise travel. Estimates indicate that 'first-time cruise customers account for a third of passengers on ocean cruises' (Cullinan, 2016, p. 4). Those unfamiliar with cruise vacations would potentially view the circumscribed environment of a cruise ship at sea with some trepidation.

Bubble trouble

That cruise-ship bubbles have potential drawbacks is no surprise to the companies that operate them. Research reported by a trade journal demonstrates a range of concerns and

> [t]he most common excuse [for not taking a cruise vacation] in a poll of non-cruisers was 'I'll be trapped with people I don't like' followed by 'I'll be bored', 'I'll get seasick', 'I'll get claustrophobic on board', and 'cruises are for old people'. (Gill, 2010, p. 22)

A cruise vacation, once it starts, cannot be readily exited. Taking such a vacation, similar to a resort enclave, is an extended commitment to an enclosed space. Passengers make choices that have consequences for themselves:

> Cruise ships may appear to be floating resorts, but you can't check out and go someplace else if you don't like your ship. Whichever one you choose will be your home for seven days or more in most cases. The ship will determine the type of accommodations you'll have, what kind of food you'll eat, what style of entertainment you'll see, and even the destinations you'll

> visit. If you don't enjoy your ship, you probably won't enjoy your cruise. (Coffman, 2014, p. 140)

The cruise-ship bubble, in addition to insulating passengers from the outside world, simultaneously 'traps' people with the co-occupants of the same enclosed space; there may be the impression that one can neither escape the ship nor the other people on board (Jaakson, 2004; Wood, 2000). Shipboard spaces are sold in a matter that demonstrates an awareness of this problem, one that is emblematic of the 'bubble trouble' marketing current.

Prospective passengers are sometimes offered advice about selecting a ship where they would not meet incompatible ship-mates. It is recommended by the author of one cruise-vacation guidebook that 'during the annual Spring Break (usually in March, but dependent on the decisions of the school-curriculum)' one should try to 'avoid the hordes of students causing mayhem' (Ward, 2017, p. 27). There are certain types of co-passengers who are perceived to be a problem within a confined space because one cannot escape the noise and nuisances they create. The cruise-ship environment is not seen as conducive to silence and serenity: '*Quiet* and *cruise* [emphasis in the original] aren't often mentioned in the same breath. The very nature of a cruise is community – upward of 3,000 customers share their vacations with perfect strangers' (Eisen, 2007b, p. 44).

Unflattering comparisons have been made between cruise ships and other environments, some of which have an association with older consumers rather than boisterous youth. These comparisons have been criticized within trade publications. One industry commentator noted that '[s]omebody actually said cruising is like being trapped for days on something that sounds like a God awful combination of Butlins [a chain of holiday camps in the United Kingdom] and a retirement home' (Hayhurst, 2008, p. 21). Selling cruise vacations, for travel agents, 'is often about correcting a dated view of cruising' (Cullinan, 2016, p. 6) that portrays passengers as having a particular age profile. One way to address this type of 'bubble trouble' – whereby cruise-ship enclaves are only thought to attract senior travelers – is to project images of cruise travel that are desirable to a broader clientele. Travel agents who sell cruise vacations play an important role in this process, a phenomenon noted by the author of a trade journal article who interviewed travel agents. In an effort to combat the perception that 'cruising is only for the elderly', one travel agent's 'top tip is to focus on the ship' (Cullinan, 2016, p. 5). This travel agent indicates that

> for a family [she] might talk about Royal Caribbean, and go through the ice skating rink, the pirate boats with water cannons, kids and teen clubs, as well as the spa for mum, and gym for dad. It can be hard to find a hotel or resort on land that ticks all the boxes for everyone, but ships are able to. (Cullinan, 2016, p. 5)

Through the advice offered by travel agents, certain messages are conveyed about insular cruise-ship environments to sought-after segments of passengers. This marketing current, one that responds to problems associated with insularity, is focused and can be directed towards particular types of consumers – including, as indicated previously, the family market.

Being co-present with other passengers – young or old – is not the only source of worry or concern associated with confinement. A belief that there may be a restricted selection of activities onboard is also a problem. The potential for passenger boredom is sometimes addressed in cruise-vacation guidebooks that feature questions or concerns frequently articulated by passengers. One such question is: 'Won't I get board?' (Ward, 2017, p. 26). In response to the possibility that passengers could become 'ship-bored' shipboard, cruise-ship companies stress the activity-filled nature of their vessels. Recently-constructed ships

> take the facilities up a notch and include waterslides, miniature golf, in-line skating tracks, rock-climbing walls, and even a surf simulator or bowling alley. For joggers there's the outdoor promenade deck or a designated track for running and walking. Children and teens have their own playrooms, swimming pools, video arcades, and, in some cases, lounges and party rooms. (Coffman, 2014, pp. 19–20)

The comprehensiveness of shipboard facilities is frequently promoted in a manner that would seem to be a pre-emptive reaction to the belief that boredom could be possible.

That cruise-ship passengers could become seasick or contract a communicable virus threatens the orchestrated nature of consumption within the cruise-ship bubble. The experience of being at sea may be perceived as pleasant isolation but it also raises concerns about seasickness. Seasickness is seen as a relatively rare condition that, if it does occur, can be cured with purchasable remedies. One estimate indicates that 'fewer than 3 percent of passengers become seasick' (Ward, 2017, p. 36). Recommended remedies include certain pharmaceutical products and 'ginger in powdered form' (Ward, 2017, p. 37). The characteristics of present-day cruise ships are also mentioned as a solution to the problem of seasickness. These ships

> are equipped with stabilizers that eliminate much of the motion responsible for seasickness. Unless your cruise includes the open sea and wind-whipped water, you may not even feel the ship's movement, particularly if your ship is a megaliner. For first-time passengers concerned with seasickness, a megaliner is precisely the ship of choice. (Coffman, 2014, p. 123)

The solution to seasickness is therefore obtainable in the marketplace, either in the form of products with medicinal qualities or through the purchase of particular types of cruise holidays – for instance, on board 'megaliners'. Travel agents are prompted to provide prospective customers, especially first-time passengers, with advice regarding seasickness and the purchase of cruise vacations:

> Eastern Mediterranean cruises, for instance, will tend to dock at a different port every day and give passengers the chance to get out and explore. It is therefore a better choice of cruise for someone concerned about sea-sickness than a trans-Atlantic voyage where much of the cruise is spent in the open sea. (Johnson, 2008, p. 28)

The 'problem' of seasickness is therefore downplayed as resolvable if the necessary steps are taken. Packaging the shipboard experience for consumers involves promoting certain actions or practices that could help make the trip more pleasant. Options are directed towards passengers who would potentially respond well to a marketing current that offers reassurance and promises to eradicate certain worries or doubts.

An unpleasant shipboard experience can also be caused by the rapid spread of a virus – for example, a norovirus. Virus outbreaks are not peculiar to cruise ships. However, industry advocates appreciate that a connection is often made between cruise travel and noroviruses. This connection has reached the point where '[n]orovirus is often termed the cruise ship virus, even though the vast majority – some 60% to 80% of outbreaks – occur on land' (Coffman, 2014, p. 126). This notion of a dedicated cruise ship viruses is disputed:

> There's no such thing as a cruise ship disease. In confined quarters, certain respiratory and gastrointestinal diseases can quickly spread through person-to-person contact – just as they do in schools, nursing homes, hospitals, and day-care centers. (Coffman, 2014, p. 124)

The metaphorical cruise-ship bubble is portrayed as comparable to other people-filled places. Its insularity, in relation to communicable viruses, is not characterized as peculiar.

The current that encapsulates these various 'troubles with bubbles' is designed to address concerns and doubts, especially those expressed by prospective passengers unfamiliar with cruise vacations. There are messages conveyed by consumer-oriented publications as well as trade journal articles that are imbued with the theme of reassurance. Travel guidebooks and magazines speak directly to cautious, possibly first-time, cruise passengers. Trade publications offer advice to those involved with the selling of cruise vacations who may encounter hesitant customers. Selling-related endeavors that are part of this metaphorical current therefore aim to address the perceived drawbacks associated with enclosure. This current tries to propel reluctant customers towards a decision to make a cruise-vacation purchase.

Conclusions and a research agenda

Packaged, insular spaces such as festival marketplaces, event centers, and sports stadiums have become a crucial instrument in the orchestrated production and promotion of post-industrial cities and towns as well as more sparsely populated places. The cruise-ship environment is also a packaged, insular domain. Packaging urban and rural areas so that they appear as arenas of consumption, and thus re-imagining and re-imaging these places, resembles the practice of projecting carefully chosen messages regarding the nature of cruise-ship environments. Spaces that promise familiarity, comfort, and abundance are also portrayed in a manner that, through the deft use of marketing, tries to offset passengers' concerns. Tourism researchers have examined the nature of bubble environments and their value as commodified spaces (Freitag, 1994; Weaver, 2005a; Wood, 2000). However, the selling of tourist bubbles – specifically, the complexities that may underpin this process – has not been explored. Bubbles at sea that are sold to tourists as providing desirable enclosure also have the potential to offer unwanted confinement. The insular nature of cruise ships provokes a host of worries. This 'containment conundrum', that cruise-ship bubbles offer both pleasurable encapsulation as well as worry-inducing confinement, prompted the idea that two currents steer the selling of the shipboard experience. Super-sized cruise ships have been the focus of this study, but smaller-scale vessels could also be examined in relation to the 'containment conundrum'.

The way in which this study is constructed has drawbacks. It addresses selling as opposed to spending and patterns of consumption. Rather than a balanced account involving both the selling and consumption of shipboard space, only a scholarly analysis of selling is undertaken. This approach, to its credit, exposes marketing strategies that play a crucial role in intensifying consumers' desires with respect to particular spaces. Visual materials such as website and brochure photographs could be examined in addition to written texts drawn from guidebooks, trade publications, and popular magazines. The perspectives and actions of consumers – not always well represented by published texts – warrant study too. Employees are worthy of similar consideration; they experience enclosure as well, but not necessarily in the same way consumers do. The diverse mix of both consumers and employees from a range of countries means that there is potential for cross-cultural research.

There is scope for future research to examine still-unexamined aspects of the cruise-ship bubble. The relative force of each of the two currents identified has not been addressed. It is possible that one current is, at times, more forceful than the other. The intensity of different currents, and perhaps their intensity varies over time, is worthy of study. Do selling-related endeavors that target the problems associated with shipboard viruses increase in volume or magnitude after a well-publicized outbreak occurs? As cruise ships become more expansive and feature more amenities and services, the two currents addressed in this paper will no doubt continue to flow in some fashion and reflect attempts by cruise-ship companies to set a course consistent with profitability.

One could narrow the focus and examine the selling of specific types of shipboard space such as cabins and suites. Convincing passengers to purchase more expensive accommodations, such as a suite with a balcony, is an important source of revenue for cruise-ship companies. Published sources – those that target both consumers and industry professionals – make reference to square footage, the place where the cabin or suite is situated on board the vessel, and the in-room amenities. What types of marketing messages characterize the selling of accommodation space? How are certain spaces, such as cabins or suites, tied to particular – and even emerging – revenue-producing activities?

There is potential to explore broader themes in relation to enclave environments. The phenomenon of enclosure cannot be separated from the existence of connection. Bounded sites, including cruise-ship enclaves, are not completely isolated from the influence of the outside world. The metaphorical currents that have been explored reflect wider connections. One current, 'bountiful bubbles', includes comparisons between cruise ships and shopping malls. It also draws attention to the familiar, branded products on board. Shipboard spaces mirror, and are thus connected to, contemporary consumer culture. The thrust of the other current, which responds to the problem classified as 'bubble trouble', is shaped by the popularity of wider travel-based traditions – such as the North American rite of 'spring break' (Monterrubio & Andriotis, 2014) – and the spread of communicable viruses, many of which are brought on board the vessel. This tension between insular, contained domains and the way in which they are connected with the outside world is worthy of more, and deeper, consideration. To what extent are cruise ships simultaneously experienced as places of enclosure and places of connection? How are these seemingly antithetical notions, enclosure and connection, intertwined within cruise-ship (or, more broadly, tourism) enclaves?

Places are ambiguous in terms of the manner in which they are experienced and, where relevant, sold. This study of insular cruise-ship spaces has demonstrated that the desirable aspects of enclosure are conspicuously promoted alongside marketing messages designed to reduce unease. There is scope for a deeper exploration of emotions tied to insular environments. Fear and worry are emotions which are sometimes associated with international travel and being away from home (Reisinger & Mavondo, 2005). There are opportunities to examine these emotions in relation to tourist bubbles as well as very different emotions – calmness, a sense of security, and desire – in the same types of spaces.

Future conceptual work with respect to the selling of insular, enclave environments in the tourism industry could investigate the commercial structures that support the shaping of societal wants as well as efforts to influence individual consumers. The theme of complementarity – prominent in this study in relation to the different ways shipboard space is sold – could be extended. Although marketing messages may be competitive between brands, marketing campaigns collectively operate in a complementary manner. People make choices that are consistent with the combined power of messages that promote purchasing. The purchasing opportunities that buttress consumer culture are potentially tied to an array of contained spaces – for example, the home, the gated community, or the shopping mall. How are consumption and commerce – almost uncontained and uncontainable within the boundaries of capitalist societies – aided by the presence and functioning of contained spaces? The drive for continued capital accumulation prompts business owners to expand their operations across different spaces (for example, countries with untapped markets) as well as into different types of spaces (including insular, revenue-creating realms).

Acknowledgement

I wish to thank Sandra Wall-Reinius, Jarkko Saarinen, John Weaver, Damian Goulbourne, and Natalie Fawcet for their assistant with this manuscript.

Disclosure statement

No potential conflict of interest was reported by the author.

ORCID

Adam Weaver http://orcid.org/0000-0002-0319-1354

References

Carpenter, M. L., & Upchurch, R. S. (2008). A five-year content analysis of articles in the American Resort Development Association's *Developments* magazine (2002–2006). *Journal of Retail & Leisure Property, 7*(2), 149–160.

Coffman, L. (2014). *Fodor's complete guide to Caribbean cruises* (5th ed.). New York, NY: Fodor's Travel.

Cohen, E., & Neal, M. (2012). A middle eastern Muslim tourist enclave in Bangkok. *Tourism Geographies, 14*(4), 570–598.

Coles, T. (2003). Urban tourism, place promotion and economic restructuring: The case of post-socialist Leipzig. *Tourism Geographies, 5*(2), 190–219.

Coomber, L., Lee, Z.-H., Lee, Y.-L., & Cairncross, G. (2006). A thematic review of Australia's *Club Management* magazine. *International Journal of Hospitality & Tourism Administration, 7*(2/3), 109–134.
Cruz, G. (2016). Disney dream: Recent makeover adds new spaces for kids and adults, all with a sprinkle of pixie dust. *Cruise Travel, 37*(5), 14–17.
Cullinan, C. (2016, May). Take to the water. *Travel Trade Gazette UK & Ireland,* pp. 4-7.
De Botton, A. (2004). *Status anxiety*. New York, NY: Vintage Books.
Duman, T., & Mattila, A. S. (2005). The role of affective factors on perceived cruise vacation value. *Tourism Management, 26*(3), 311–323.
Eisen, D. (2007a). Attracting first-time cruisers. *Travel Agent, 329*(2), 20–22.
Eisen, D. (2007b). The sounds of silence. *Home-Based Travel Agent, 2*(11), 44–45.
Eisen, D. (2008). First takes: Novice cruisers offer special challenges for agents. *Travel Agent, 332*(13), 54–57.
Freitag, T. (1994). Enclave tourism development: For whom the benefits roll? *Annals of Tourism Research, 21*(3), 538–554.
Gill, R. (2010, March). News in brief. *Travel Trade Gazette UK & Ireland,* p. 22.
Goffman, E. (1959). *The presentation of self in everyday life*. New York, NY: Doubleday Anchor.
Gutberlet, M. (2016). Socio-cultural impacts of large-scale cruise tourism in Souq Mutrah, Sultanate of Oman. *Fennia, 194*(1), 46–63.
Hall, C. M. (2008). Servicescapes, designscapes, branding, and the creation of place-identity: South of Litchfield, Christchurch. *Journal of Travel & Tourism Marketing, 25*(3–4), 233–250.
Hayhurst, L. (2008, May). Breaking into cruise. *Travel Trade Gazette UK & Ireland*, pp. 21-22.
Horowitz, D. (2004). *The anxieties of affluence: Critiques of American consumer culture, 1939–1979*. Amherst, MA: University of Massachusetts Press.
Horowitz, D. (2012). *Consuming pleasures: Intellectuals and popular culture in the postwar world*. Philadelphia, PA: University of Pennsylvania Press.
Iaquinto, B. L. (2011). Fear of a *lonely planet*: Author anxieties and the mainstreaming of a guidebook. *Current Issues in Tourism, 14*(8), 705–723
Jaakson, R. (2004). Beyond the tourist bubble? Cruiseship passengers in port. *Annals of Tourism Research, 31*(1), 44–60.
Jackson, P. (2010). Food stories: Consumption in an age of anxiety. *Cultural Geographies, 17*(2), 147–165.
Jacobsen, J. K. S. (2003). The tourist bubble and the Europeanisation of holiday travel. *Journal of Tourism and Cultural Change, 1*(1), 71–87.
Johnson, C. (2008, May 2). Putting wind in the sales. *Travel Weekly Australia,* p. 28.
Kearns, R., & Philo, C. (Eds.). (1993). *Selling places: The city as cultural capital, past and present*. Oxford: Pergamon Press.
Kirillova, K., Lehto, X. Y., & Cai, L. (2017). Existential authenticity and anxiety as outcomes: The tourist in the experience economy. *International Journal of Tourism Research, 19*(1), 13–26.
Knego, P. (2015). *Regal princess*: Princess Cruises' newest and largest vessel made a star-studded American debut. *Cruise Travel, 36*(6), 26–31.
Knudsen, D. C., Soper, A. K., & Metro-Roland, M. (2007). Commentary: Gazing, performing and reading: A landscape approach to understanding meaning in tourism theory. *Tourism Geographies, 9*(3), 227–233.
Kolberg, S. (2013). Constructing a 'democratic' dreamworld: Carnival cruise ships and an aesthetic of optimism. *Journal of Consumer Culture, 16*(1), 3–21.
Kwortnik, R. J. (2008). Shipscape influence on the leisure cruise experience. *International Journal of Culture, Tourism and Hospitality Research, 2*(4), 289–311.
Larsen, S., Wolff, K., Marnburg, E., & Øgaard, T. (2013). Belly full, purse closed: Cruise line passengers' expenditure. *Tourism Management Perspectives, 6*, 142–148.
Leppert, J. (2016a). *Harmony of the seas*: The world's largest cruise liner melds the best of RCI's oasis- and quantum-class vessels. *Cruise Travel, 38*(3), 16–20.
Leppert, J. (2016b). *Norwegian escape*: NCL's newest – And largest – Megaliner expands and enhances the breakaway class. *Cruise Travel, 38*(1), 26–31.
Lukas, S. A. (2008). *Theme park*. London: Reaktion Books.

Marchand, R. (1985). *Advertising the American Dream: Making way for modernity, 1920-1940*. Oakland, CA: University of California Press.

Miles, S. (2010). *Spaces for consumption, pleasure and placelessness in the post-industrial city*. London: Sage.

Minca, C. (2000). 'The Bali Syndrome': The explosion and implosion of 'exotic' tourist space. *Tourism Geographies, 2*(4), 389–403.

Monterrubio, J. C., & Andriotis, K. (2014). Social representations and community attitudes towards spring breakers. *Tourism Geographies, 16*(2), 288–302.

Nunkoo, R., & Ramkissoon, H. (2016). Stakeholders' views of enclave tourism: A grounded theory approach. *Journal of Hospitality & Tourism Research, 40*(5), 557–588.

O'Barr, W. (1994). *Culture and the ad: Exploring otherness in the world of advertising*. Boulder, CO: Westview Press.

Park, S., Ok, C., & Chae, B. (2016). Using Twitter data for cruise tourism marketing and research. *Journal of Travel and Tourism Marketing, 33*(6), 885–898.

Peel, V., Sørensen, A., & Steen, A. (2012). 'Unfriendly, unfunny, and tyrannical': An exploratory study of the travel guidebook in the Australian print media. *Tourism Analysis, 17*(3), 299–309.

Petrick, J. F. (2004). First timers' and repeaters' perceived value. *Journal of Travel Research, 43*(1), 29–38.

Reisinger, Y., & Mavondo, F. (2005). Travel anxiety and intentions to travel internationally: Implications of travel risk perception. *Journal of Travel Research, 43*(3), 212–225.

Saarinen, J. (2017). Enclavic tourism spaces: Territorialization and bordering in tourism destination development and planning. *Tourism Geographies, 19*(3), 425–437.

Schöllmann, A., Perkins, H. C., & Moore, K. (2001). Rhetoric, claims making and conflict in touristic place promotion: The case of central Christchurch, New Zealand. *Tourism Geographies, 3*(3), 300–325.

Seldon, C., & Seldon, L. (2013). *Royal princess*: New diversions abound on Princess Cruises' biggest 'Love Boat' ever. *Cruise Travel, 35*(3), 12–14.

Smith, A. (2005). Conceptualizing city image change: The 're-imaging' of Barcelona. *Tourism Geographies, 7*(4), 398–423.

Su, X. (2010). The imagination of place and tourism consumption: A case study of Lijiang Ancient Town, China. *Tourism Geographies, 12*(3), 412–434.

Vogel, M. P., & Oschmann, C. (2013). Cruising through liquid modernity. *Tourist Studies, 13*(1), 62–80.

Volo, S., & Pardew, D. L. (2013). The Costa Concordia and similar tragic events: The mathematics and psychology of the loss and the restoration of travellers' trust. *Current Issues in Tourism, 16*(2), 197–202.

Ward, S. V. (1998). *Selling places: The marketing and promotion of towns and cities 1850–2000*. London: Routledge.

Ward, D. (2017). *Berlitz cruising and cruise ships. 2017*. London: Berlitz Publishing.

Warde, A. (2004). Consumption, identity-formation and uncertainty. *Sociology, 28*(4), 877–898.

Weaver, A. (2005a). Spaces of containment and revenue capture: 'Super-sized' cruise ships as mobile tourism enclaves. *Tourism Geographies, 7*(2), 165–184.

Weaver, A. (2005b). The McDonaldization thesis and cruise tourism. *Annals of Tourism Research, 32*(2), 346–366.

Wood, R. (1994). Hotel culture and social control. *Annals of Tourism Research, 21*(1), 65–80.

Wood, R. (2000). Caribbean cruise tourism: Globalization at sea. *Annals of Tourism Research, 27*(2), 345–370.

Yan, G., & Santos, C. A. (2009). 'China, forever': Tourism discourse and self-Orientalism. *Annals of Tourism Research, 36*(2), 295–315.

Yarnal, C. M. (2004). Missing the boat? A playfully serious look at a group cruise tour experience. *Leisure Sciences, 26*(4), 349–372.

Yarnal, C. M., & Kerstetter, D. (2005). Casting off: An exploration of cruise ship space, group tour behaviour, and social interation. *Journal of Travel Research, 43*(4), 368–379.

Young, S. J. (2015). First-time cruisers. *Travel Agent, 346*(13), 14–16.

Cruisers in the City of Helsinki: staging the mobility of cruise passengers

Kaisa Paananen and Paola Minoia

ABSTRACT
International cruise passengers moving in urban destinations have particular time limitation that can make enabling and disabling elements of mobility meaningful on the quality of their visit. Identifying these elements is essential to improve their independent movement. Based on a staging mobilities framework that considers the dimensions of physical settings, material spaces, design (PMD), social interactions (SI) and embodied performances (EP) *in situ*, the research has analysed location-specific information in Helsinki, a popular port of call in the Baltic Sea. The study has used go-along observations and mobile application in data collection. The produced dataset was analysed by combining GIS-methods and content analyses. As a result, five categories were identified under the dimension of PMD: wayfinding tools, unexpected situations, lack of rest spots and walkability. Categories identified under the dimension of SI were local people, service providers and travel companions, and other tourists. Categories identified under the dimension of EP were traffic behaviour, occasioned activities, sense of direction, planning, and time-related anxiety. Urban destinations such as Helsinki can apply these results in practice, to make the movement of independently moving cruise passengers as effortless as possible, contributing to a better experience of the city space for both tourists and other city users. The proposed methodology could also be used to analyse other mobility-related phenomena.

1. Introduction

'*Do we have time to visit the church? What if we miss the ship?* 'These worries expressed by cruise passengers while moving independently from the port terminal of Helsinki to the city centre, represent their sense of insecurity while moving out of the tourist bubble of the ship (Judd, 1999). Cruise passengers, a category of organized mass tourists (Diedrich, 2010), base their feeling of safety from the controlled environment of the ship, where personnel and activities are at their service to fulfill their needs (Jaakson, 2004). When a ship docks on a port of call, passengers have three mobility possibilities: to remain on board inside the safe bubble, to join activities organized by the cruise in the arrival place, or to visit the place independently. The moment of disembarkment at the port is, therefore, quite exciting; often, these daily tourists arrive unprepared to the port of call, and their mental map of the destination is rather uncertain (Aranburu, Plaza, & Esteban, 2016; Lew & McKercher, 2006; Lynch, 1960).

In Helsinki, out of the 96% of passengers who depart the cruise, about 60% take organized tours, and only 35% move independently (G. P. Wild (International) Limited & Business Research & Economic, 2015). Helsinki's local government, like many other coastal locations, welcomes the city's inclusion in international cruise itineraries because of the positive expectation about economic benefits brought by the disembarking tourists as consumers of goods and users of local transport and other urban services.

The focus of this study is on independent cruise passengers moving inside the destination by walking or using public transportation and aims to shed light on two questions, one substantial and one methodological. The first research question regards the type of urban factors either facilitating or challenging the walking mobility of cruise passengers in the city; the second question looks at the applicability of go-along methods for spatial analysis of this tourist mobility. The research is based on an empirical study of Helsinki, a port of call specialized on cruise ships, and where passengers stop to spend only a few hours in a daytime (Marti, 1990).

Cruise tourism started in the early 1960s and increased rapidly worldwide, growing from 3.8 million passengers in 1990 to 24.2 million in 2016 (Cruise Market Watch, 2016), with estimated 25.3 million passengers for the year 2017 (Cruise Lines International Association (CLIA), 2017).

Earlier studies (De Cantis, Ferrante, Kahani, & Shoval, 2016; Ferrante, De Cantis, & Shoval, 2016) have demonstrated the relevance of research on cruise passengers' behaviour at the destinations, to ensure good management and provision of the necessary services, and guarantee positive tourist experiences. Tourist usually select the cruises based on ports of call and their attractiveness, and decreases in satisfaction's rates can make their inclusion in cruise tours at risk (Henthorne, 2000) . Moving from these considerations, our study would like to contribute to the analytical understanding of spatio-temporal data on tourist mobility integrated with qualitative information, to evaluate cruise passengers needs and potentially address problems of access to, and satisfactorily visit of the city.

2. Background studies and theoretical framework

Cruise ship tourism is still of minor interest academically, despite its significant growth and impacts on coastal cities (De Cantis et al., 2016; Ferrante et al., 2016; Papathanassis & Beckmann, 2011; Rodrigue & Notteboom, 2013; Xie, Kerstetter, & Mattila, 2012). Previous studies have looked particularly at economic aspects, for instance on produced income at the destinations (Brida, Pulina, Riaño, & Zapata-Aguirre, 2012; Penco & Di Vaio, 2014; Satta, Parola, Penco, & Persico, 2015); passenger satisfaction (Satta et al., 2015; Silvestre, Santos, & Ramalho 2008); preference choices of destinations, returns and recommendations (Andriotis & Agiomirgianakis, 2010; Baker & Fulford, 2016; Larsen & Wolff, 2016); or produced environmental and socio-economic impacts, also negative (Brida & Zapata, 2010; Carić, 2016; Jordan & Vogt, 2017).

Other researches on ports of call have focused on elements contributing to their quality, for instance in terms of natural and cultural assets, port facilities, security, costs and marketing (Manning, 2006); quality of services in the cities, sea, land and air connections (McCalla, 1998); proximity to tourist attractions attractiveness in terms of infrastructures, competitive prices, products and services, with emphasis on market segmentation to satisfy different categories of customers (De Cantis et al., 2016). All these studies have highlighted various structural conditions of mobility of cruise passengers, while behavioural issues in their short visits have been mostly neglected (De Cantis et al., 2016; Ferrante et al., 2016; Xie , Kerstetter, & Mattila 2012;Yarnal & Kerstetter (2005)). A qualitative study proposed by Andriotis and Agiomirgianakis (2010) about Heraklion, Crete, has shown practical problems encountered by independent tourists even to simply hiring a taxi or finding the main tourist attractions. De Cantis et al. (2016) and Ferrante et al. (2016) have made use of GPS devices to study cruise tourist mobility with the aim to, respectively, identify passenger profiles based on their behaviour at the destination, and to create a database on cruise passengers mobility onshore. Different types of mobile methods for data collection have become popular in tourism studies, using new technology; GPS tracking and recording itineraries (Edwards & Griffin, 2013, McKercher, Shoval, Ng, & Birenboimim, 2012; Shoval & Ahas, 2016; Xie, Kerstetter, & Mattila 2012 Shoval & Isaacson, 2007), also allowing better understanding of tourist activities and supporting specific travel behaviours (Zheng, Huang, & Li, 2017). Moreover, passive mobile positioning has been used for collecting big data, e.g. from tourism statistics and geocoded photos (Ahas, Aasa, Mark, Pae, & Kull, 2007).

Smartphones and tablets are used to trace locations by GPS, cell-tower identification and Wi-Fi positioning. Even though the development of mobile applications in smartphones has not been used extensively so far, there is still good potential also for enriching data with qualitative information related to the visiting experiences (Shoval & Ahas, 2016; Kang 2016). Furthermore, various technological applications are being developed that are not yet reported in tourism literature.

Our study has used Jensen's *staging mobilities* framework, an analytical tool that has inspired a newly emerging field named *mobilities design* focusing on physical interventions and design-related decisions that can enable or disable mobile situations (Jensen, 2013, 2014a, b, Veijola & Falin, 2016). The *staging mobilities* framework is located in the *new mobilities paradigm* (Cresswell, 2006; Lanng & Jensen 2016; Sheller & Urry, 2006; Urry, 2000, 2010;), looking at mobility as an integral part of everyday life. Mobilities are understood in their socio-material configurations (Fullagar, Wilson, & Markwell, 2012; Larsen, 2001; Tranberg Jensen, Gyimothy, & Jensen, 2015) that are significant in individual social and emotional lives (Urry, 2000)

Jensen's *staging mobilities* model looks at two levels of staging: from *above*, regulated by planning, design, regulations, and institutions; and from *below*, produced by interaction and individual performances. The two levels construct the mobile situation within three dimensions: (1) Physical settings, Material space, and Design (PMD); (2) Social Interaction (SI) and (3) Embodied Performances (EP). In the first dimension, PMD, individuals move and use roads, signs, traffic lights, etc. that are both designed and regulated to facilitate mobility and its speed (Sheller 2017). According to Bissell (2010) mobility also makes use of ordinary materials like laptops, mobile phones, and tickets, and therefore, PMD could also include them. These elements do not only occupy the background but are dynamic formations as *mobile biotopes*: *'Manmade environments where mobility and living are ongoing and constant'*(Jensen, 2013, p. 47–48). Human bodies interact with these environments intended as semiotic systems that can make sense of the mobile situations. Material spaces form the landscapes that are organized from *above* and at the same time, staged from *below*, through gestures and bodily postures performed by people. Mobility '*affordance* ' (*ibid*.: pp. 94–95) describes the relationship of the moving body with the material environment that either opens up or narrows down the modes of travell. Mobility *aesthetics* (Jensen, 2014a, pp. 48–49) reveals the subjectivity produced by mobile bodies that are affected by particular rationalities, cultural norms, and psychologies. In fact, different kinds of effects like stress and irritation produce an impact on the travel atmosphere (Bissell, 2010) and, consequently on the very act of mobility.

Social interaction (SI) is crucial for understanding mobilities *in situ* (Jensen, 2013, p. 65). Tourists move within the city with family, friends, partners, groups and with other companies met for example during the cruise or on the street ('*temporary congregations*'), therefore being '*mobile withs*' *together*. Interactions between mobile bodies may seem random, but constitute a process of negotiation that takes place while in motion ('*negotiation in motion*': Jensen, 2014a, p. 46): this is rather common in the case of tourists moving in unfamiliar environments. The last dimension, Embedded Performances (EP) explains how the body influences mobility patterns, its movements and 'the *mobile self* ' (Jensen, 2014b).

2.1. Study context

The case study focuses on the port of call of Helsinki, located on the northern shore of the Baltic Sea. International cruising in this area is seasonal, usually happening during the summer months. Figure 1 represents the top cruising destinations in the area, namely: Copenhagen, Stockholm, Rostock, St. Petersburg, Tallinn, Helsinki and Kiel.

The number of passengers has grown in the Baltic region, and more and more visitors disembark in Helsinki. Figure 2 shows this growing trend. In summer 2016, Helsinki was visited by 409,000 passengers coming from 240 ships (Port of Helsinki,

2017) and originally, mostly from Germany, United States, UK, Spain and Canada (Visit Helsinki, 2017). In 2014, the average time spent in Helsinki was 4.5 h, while half of the passengers stayed 2–3 h onshore; 87% of the passengers were first-time visitors in the city. The average size of the party was two persons, and the average age was 58 years: 48% were 65 years or older, and only 6% were under the age of 25 (G.P. Wild 2015).

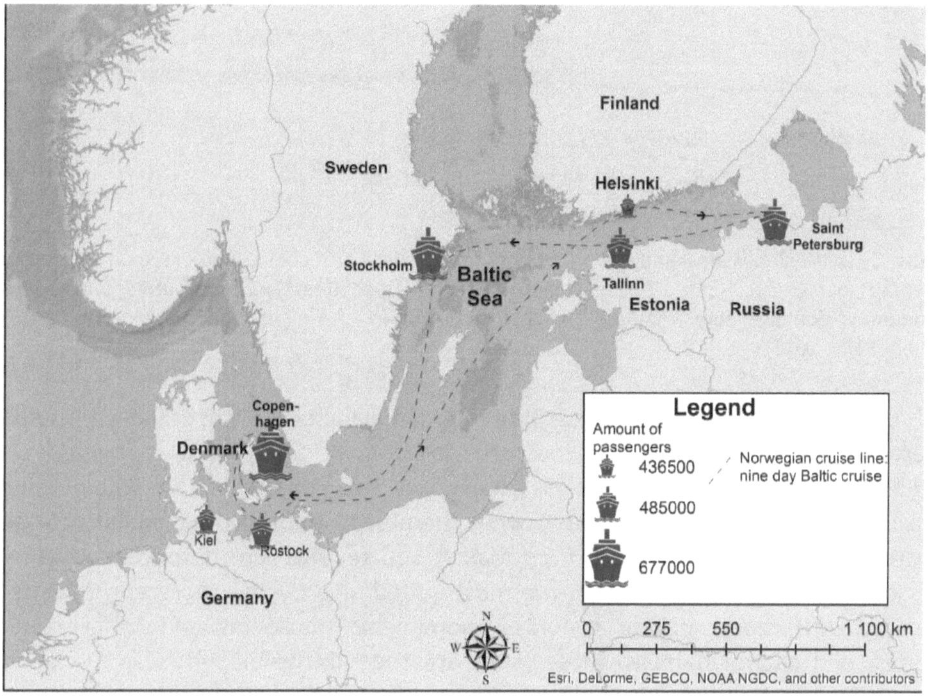

Figure 1. Ports of calls at the Baltic Sea region with most international cruise ship passengers and an example route of one Baltic Sea cruise, the nine days Baltic Cruise of Norwegian Cruise Lines (Authors' own production). Basemap source: Esri, DeLorne, GEBCO, NOAA, NGDC and other contributors.

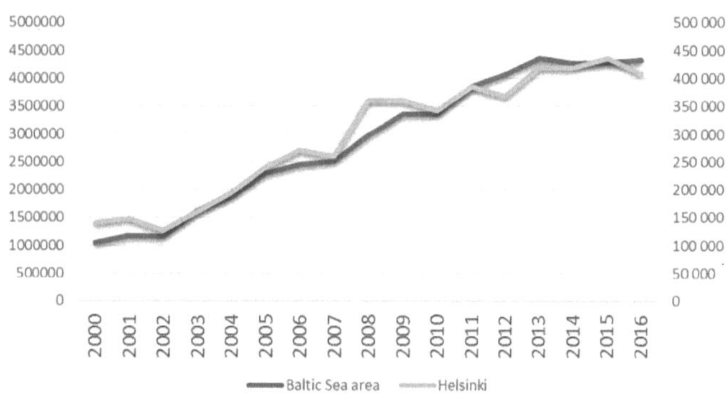

Figure 2. The number of passengers in Helsinki and in the Baltic Sea area in 2000–2016. The scale on the left shows the changes in the whole Baltic Sea area. The scale on the right shows the changes in Helsinki. (Authors' own production). Data source: Cruise Baltic, 2016

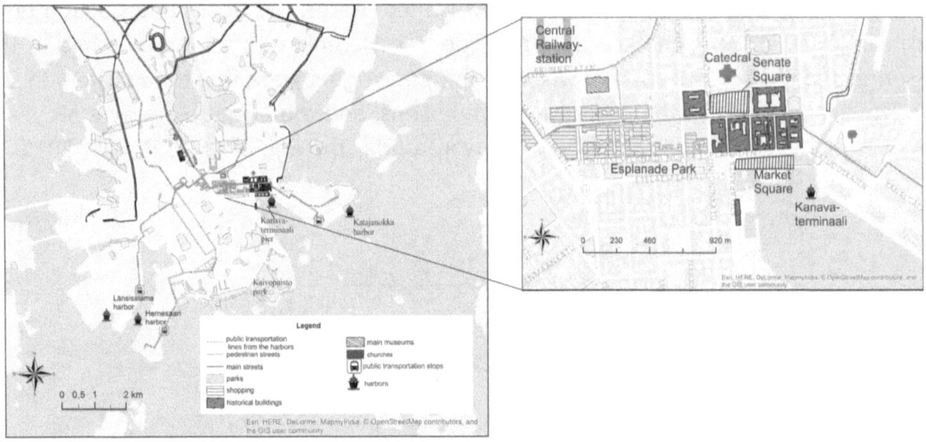

Figure 3. Helsinki city area and the most important tourist attractions. Authors' own production). Basemap source: Esri, HERE, DeLorne, MapmyIndia, © OpenStreetMap contributors, and GIS user community; Contains data from the National Land Survey of Finland Topographic Database 06/ 2014; © HSL 2016).

Helsinki city has four harbours serving international cruises: Hernesaari, Länsisatama (Western harbour), Eteläsatama (Southern Harbour, including Kanavaterminaali pier) and Katajanokka. As Figure 3 shows, Katajanokka and Eteläsatama are within a short walking distance to Market Square, while Hernesaari and Länsisatama are located, respectively, 4.5 and 4 km away from Market Square. The core historical centre near the seashore and the Cathedral is the most visited area by the short-term tourists in the city. Attractions include historical monuments, museums and other cultural centres, and shopping malls; these latter are specially developed near the Central Railway Station, Mannerheim street, and Kamppi area.

3. Methodology

Our study has tested a go-along observation as a data collection method with a mobile application that has allowed recording of qualitative in-depth and place-specific data (Finlay & Bowman, 2017), with information gathered while visitors were actually moving around (Kusenbach, 2003). This type of qualitative data collection's technique has been used surprisingly rarely in tourism movement studies, although the spatial information produced by tracking and recording tools could also be useful for urban tourism planning. So far, studies influenced by the mobilities' theory have rather used other sorts of qualitative ethnographic field methods, especially embodied and classic ethnographic mapping (Jensen, 2013, 2014a). As said, our go-along method made it possible to understand certain phenomena at the same time, place and space of their happening. We used GPS for geo-referencing data. Conversations between the researcher and the participants in situ could enrich the geographic data giving qualitative value and facilitating their adjustment (as indicated by Kusenbach, 2012).

Our observations were conducted with passengers that arrived at all the four harbours of Helsinki. Sampling was discretionary: the criteria for informants' selection were chosen beforehand, based on the basic demographic profile of a typical cruise

Table 1. Study participants.

Tour month		Arrival harbour		Group size		Travel company		Age		Nationality	
May	5	Hernesaari	11	One	2	Alone	2	Under 15	6	UK	12
June	5	Kanavaterminaali	10	Two	12	Family (with children)	10	15–24	7	USA	9
July	11	Länsisatama	10	Three	4	Family (without children)	4	35–44	4	Germany	6
August	11	Katajanokka	4	Four	11	Spouse	15	45–54	12	Mexico	2
September	3			Over four	6	Friends	2	55–64	8	Spain	2
						Co-workers	2	Over 64	12	France	1
									12	The Netherlands	1

passenger in Helsinki, e.g. nationality, age, and size of the travel group. Informants were approached at their departure from the ship, and the contact time lasted for about one hour during their way into the city. Tourists were asked to report any enabling and disabling element of mobility along their way. Informants paid particular attention to physical features, which were easier to detect and express. Other non-verbal factors regarding social interaction and embedded issues remained mostly unexpressed verbally but still visible and noted by the researcher. The accompanying researcher tried not to affect the route or to give any route suggestions. Observations were recorded with a mobile application that provided space – and time-related information and completed by observation reports written after each tour. The Android operating system mobile application, called Tourist Experience Tracker was designed for this study and now it can be freely downloaded from Google Play Store. We planned the design and content, and the coding was done by a computer science professional, Charlie Hornsby. The application consists of a tracker and a possibility to add comments as well as photos and includes a separate function for adding background information. The application has a feature that gives the possibility to press 'thumbs down' when the observed issue affected negatively on the movement and 'thumbs up' when it affected positively. The tracking function is based on cell-tower identification and Wi-Fi positioning.

Some methodological challenges regarded the researcher-informants relation during the exercise. Like in qualitative studies in general, the subjectivity of both researchers and research has to be considered. For us, it was essential to develop a positive connection with the informants from the very beginning, even though the good relationship always activated more interest by the passengers to discuss other issues, which interfered with the exercise. Getting people to participate proved to be overall smooth, but approaches were easier when we knew the languages spoken by the tourists. In some cases, we realized that our presence decreased the anxiety of the participants while moving into the city. Practical issues such as weather affected the go along exercise although only during one tour it was impossible to continue.

Altogether, we conducted 35 observations were involving 61 persons, between May and September 2016. Table 1 presents some characteristics of the study participants.

For the analysis of data, the primary focus was to detect enabling and disabling elements influencing the tourists' mobility towards the city, and within the city centre in the short time span they had. All data included the researchers' notes about the tourists' experiences, including quotes, body language and other expressions of tourists' behaviours while moving; clearly, then, the notes reflect the researchers' understanding about the situations. Data were georeferenced, so that all notes had their precise locations.

The data was stored in Google Developers Console and then processed with Atlas.ti-programme for content analysis. The coding was derived from the *staging dimensions* theory: PMD/enabling or disabling, SI/enabling or disabling, and EP/enabling or disabling. Altogether 271 elements were detected and coded, resulting in 150 disabling elements and 29 enabling elements within the category of PMD; 5 enabling and 23 disabling elements within the category of SI; and 26 disabling and 17 enabling elements under the category of EP.

All the comments under each code were analysed and clustered within specific categories. The space-time information was in *json* form, and through the program

Python, *json* form was coded into a shapefile able to transform the data into ArcMap. By comparing the routes and time/space information of the comments, the location of the comment could be placed on the map. Then the comments could be classified in ArcMap based on the categories created in the earlier phase.

3.1. Results: disabling and enabling categories in cruise passengers' mobility

The analysis reveals the distribution of the elements PMD, SI and EP along the routes taken by the tourists in the different areas of Helsinki. Most comments were expressed in the core tourist area, where most of the passengers headed during the first hour of their visit. Another area that was pointed by many comments was the Rock Church, reached by bus from Hernesaari harbour. The following sections describe these elements more in detail. Some specific situations are introduced through quotes taken from the go-along notes.

Figure 4 represents Jensen's framework applied to the situation of mobility in Helsinki, after the result given by the content analysis. What needs to be acknowledged is that this model is a simplification of the reality, since enabling or disabling categories rarely work on their own but are usually intertwined and occurring in the same space and time.

3.1.1. PMD-related disabling and enabling categories

Data were combined into the following categories: wayfinding tools, unexpected obstacles, rest spots and walkability. The PMD dimension was the richest one compared to SI and EP, for its elements were immediately noticeable and easy to describe.

Wayfinding tools.

> The informants are leaving the Rock Church. Next destination is the Market Square. They are looking at the map outside the church for a long time. Along the way, they need to check the route many times, and they are taking a longer route. (Notes taken while following a British group of friends of 50–55 years).

Wayfinding, in general, can be seen as a semiotic dimension, as individuals learn to read and use available signs. These tools help form mind maps more as they link different areas and landmarks in a logical way (Chang, 2015). For passengers arriving at the harbours, directional signs were among the most convenient mobility tools. They

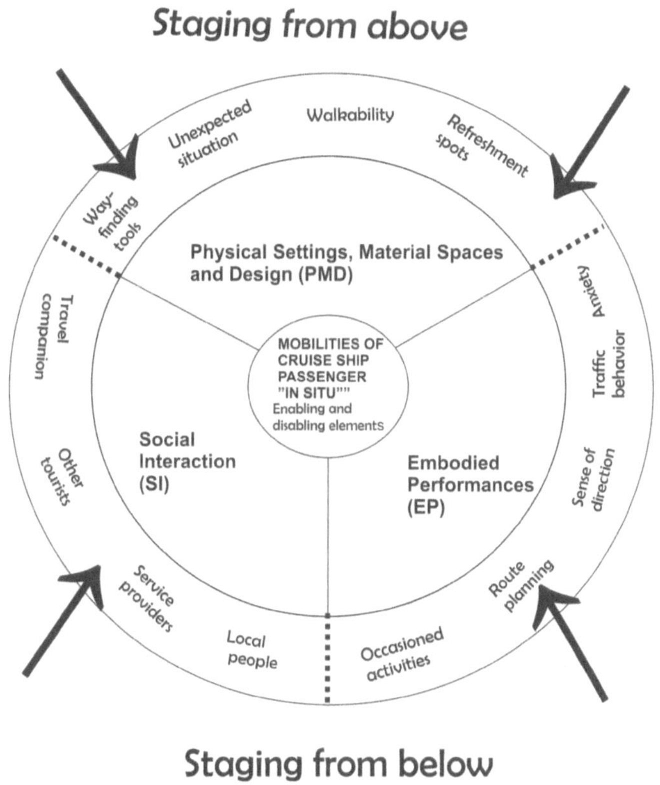

Figure 4. Staging mobility of cruise ship passengers in Helsinki, based on the staging mobilities framework. Source: Jensen (2013). (Authors' own production).

were especially important for passengers walking from Länsisatama and Hernesaari harbours. The location of the signs was interesting in the case of Hernesaari harbour, as they were located along a road that was not used by passengers, who were instead taking shortcuts through Kaivopuisto park; thus, those signs installed by city planning units (*above staging*) were missed by most visitors (*below staging*). Unfortunately, Kaivopuisto does not have directional signs, and therefore, tourists had to stop many times to read their maps. Also in the case of the Rock Church, disabling elements were mostly related to the lack of directional signs leading to the central tourist area. On the contrary, near the Market Square, lack of signs did not constitute a disabling element since many of the passengers used instead monuments as wayfinding tools. In Länsisatama harbour, directional-map stands proved to be enabling elements helping passengers to find their way more easily.

Wayfinding tools are vital for the use of public transportation, although as observed by Hall, Le-Klähn, and Ram (2017), tourists often feel unsecure about the reliability of the information given. In Hernesaari harbour, finding the bus stop turned out to be difficult because of the presence of three bus stops next to each other. Inside the public transports, the main issue was in finding the right stop to get off. Some passengers were hoping for announcements or screen information about the attractions near the stops. After departing the vehicle, directional signs to the main tourist attractions would have been helpful, while passengers were instead disoriented for the long time required to find their location on the map. Using mobile devices as wayfinding tools were much appreciated while using public transportation. Often one person in the group was responsible for following the route on a map-based application in their mobile phones. On the other hand, walking passengers relied more on traditional maps.

Unexpected situations.

> The family is taking a sightseeing tram. All of a sudden the tram stops and then goes on a special route. No information was given in English, and the family is getting nervous. (Notes taken while following a family with mother, grandmother and two boys under 25 from Germany).

While moving, unexpected situations worked as disabling elements. Examples are in the construction sites located in both Länsisatama and Kanavaterminaali harbours. These places were the first encounter that passengers had with Helsinki, and the situations there required a lot of straying. Occasional breakdowns in public transportation are also within this category; they may cause anxiety, especially when the information is only provided in Finnish. Such situations miss the control of *above* staging, as they are unplanned and very rarely local administrations intervene promptly.

Rest spots.

> A couple arrives at the Länsisatama harbour and walks from there to the tourist area. In the middle of the walk, the man needs to use the toilet. They cannot find public toilets anywhere. They need to walk all the way to the Stockman department store, and the walk is disturbed by much anxiety. (Notes taken while following a couple over 65 from the US).

The lack of free toilets along the long routes from the Hernesaari and Länsisatama harbours are surely disabling elements, as it is the lack of refreshment spots. Most passengers walking to the tourist area and especially the Market Square did not care to have coffee etc. since they had just left the ship; yet, some needed to rest, but the area did not offer free facilities.

Walkability.

> The group is taking pictures of the Uspenski Cathedral, but then a woman steps on a hole on the street that she had not noticed. She almost falls and feels strong pain in her hip where she had a surgery a few months ago. (Notes taken while following a British group of friends of 50–55 years).

Walkability is a wide concept with different definitions (Forsyth, 2015; Hall, Ram, & Shoval, 2018). According to Southworth (2005), it includes a secure network of connections to different destinations, within a reasonable time and effort and at the same time providing a pleasant and interesting context. Tourism-related walkability can be characterized by comfort, tourists appeal, activity potential, exploration (Ram & Hall, 2018; Samarasekara, Fukahori, & Kubota, 2011; Ujang & Muslim, 2014); however, in this study, the concept of walkability refers especially to the fruition of landscapes and functional infrastructure for mobility.

There were 11 enabling comments detected about walkability. One of the best cases in Helsinki is represented by the walking route from Hernesaari harbour along the seaside. However, the mobility aesthetic, its enjoyment and possible distress do not only depend on objective material factors (*above* staging level) but also on subjectivity factors like tourists' previous experiences,the cultural meaning given to observed elements, and behavioural patterns (*below* staging level). Some problems in the spatial and infrastructural conditions were seen as the cause of small accidents particularly near tourist attractions, where passengers concentrated more on tourism gazing than on looking at their pathways. Sometimes, tripping occurred in uneven paths, and incautiously even on bike lanes, with the risk of unexpected collision. This risk was higher especially around Market Square, where despite the relatively high traffic, inconspicuous marking of bike lanes and parking made pedestrians take wrong paths, which were understood as the most direct ones to get into the square.

3.1.2. SI-related disabling and enabling categories

Within the social interaction, also in line with the three categories identified by Pearce (2005), we recognized the following elements: service providers, local people, travel companions and other tourists.

Service providers.

> The family is traveling from Hernesaari harbour to the Rock Church. The bus driver does not speak English and is not able to sell them daily tickets. The family is confused, and the driver makes hand signals just to sit down. (Notes taken while following a Russian family composed of a 50–55 years old couple and four children under 25).

While using public transportation, the interaction between passengers and the driver was seen as an essential element of the *negotiation in motion*. The driver's response was based on language skills and helpfulness attitude. However, even though the situation suggested that the main driver was operating at the *below* staging, some conditions were given from *above*, since both the ships and public transport had to respect predetermined timetables, and this produced anxiety on both parties and pressure in the interaction. Some disabling elements were evident at the stops in the harbours. Although immobile, they are seen as *temporary congregations* (Jensen, 2013) since it is from those places that movements begin.

Local people.

> The party is taking a tram from the harbour towards the Senate Square. They start to talk with a local man. He gives them advice on what to see and where to get off the tram. (Notes taken while following two American couples, over 60 years old).

Interactions that passengers have with other city users is more random and staged from *below* because of its unplanned nature since it is only produced if the parties want to. In some streets outside the tourist area, locals saw tourists looking into their maps and offered help. This is a significant interaction, and apart from cases when bad advice is given, it is enabling even when informational signs are present nearby because it provides help that is considered faster and more reliable by tourists.

Some negative confrontation with the locals occurred, when cruise passengers were walking too widely on the paths or blocking the routes, like in the case of the route from Hernesaari harbour that is a popular recreational route for locals. Additionally, as already reported, bikers and passengers had confrontations near Uspenski Cathedral, where the same path has to be used by both pedestrians and bikers. These situations are emphasized when locals and passengers show different traffic behaviour.

Travel companions.

> Children are looking bored after the visit to the Rock Church. While walking towards the historic city, they see a statue in front of the natural history museum. They want to visit the museum. One of the men says that there is no time. They continue to the historical city, and children are getting even more bored. (Notes taken while following a party with two couples, grandparents and four children under 12 years old from Germany).

During the observations, it was noticed that the category of travel companions worked as both enabling and disabling element. Earlier studies (Larsen, Axhausen, & Urry, 2006) have shown that trips are used also as a way to spend time within families and close friends since, in ordinary life, free time is lacking.

Cruise ships offer opportunities for family travels, with grown-up children and older parents. Often, adult children take responsibility and thus, constitute an enabling element of their parents' mobility. However, disabling elements may take place inside families, especially when groups include old people or small children. This was particularly obvious during the long walks from Hernesaari and Länsisatama harbours. Some older group members were walking at a slower pace, and this either annoyed the younger members or create anxiety especially in situations of time constraints. Generational problems were observed when children were less interested in historical buildings while, instead, needed other leisure activities. Along with Jensen (2013, p. 83) we also noted ongoing negotiation resulting in arguments and fights about where *mobile withs*, i.e. group of passengers moving together, should go.

Other tourists.

> Informants are walking from Hernesaari harbour and are following a British party. In Kaivopuisto Park, the Brits turn into the park. The Dutch first follow them, but then start to hesitate and look at the map. Then they decide to take another route. (Notes taken while following a young Dutch couple, 25–30 years old.

On a cruise, passengers often befriend other passengers they do not know beforehand (Teye & Leclerc, 1998); but during short onshore times, arguments can be triggered by disabling factors. In our observation, this category was mentioned seven times as a disabling and eight times as an enabling element. While departing the ship in Helsinki, cruise-passengers often followed other passengers, even if these ones were not sure about the way. In some cases, different groups started exchanging opinions about routes, then going together at least for part of their journey to the city. This process could have both enabling and disabling elements.

3.1.3. EP-related disabling and enabling categories

This section includes embodied performances factors that were mostly combined with others, already introduced. Identified elements were traffic behaviour, occasioned activities, sense of direction, travel planning and anxiety. Enabling elements, especially positively unexpected, were observed mostly on the seaside path between the major tourist area and Hernesaari harbour, while disabling elements were mostly concentrated in the city centre, and sometimes in Länsisatama harbour.

Traffic behaviour.

> The party includes a woman with a child in a stroller. They try to get into a tram from the first door. The driver does not let them in and shows the right door to use. The party buys a tram ticket for the woman, even though public transportation is free for a person traveling with a child in a stroller. (Notes taken while following a Spanish family, grandparents, man, and woman with three children under 15).

Often, and especially near Market Square, where there is heavy traffic, tourists showed bad behavior creating risky situations, such as jaywalking. This is an obvious case of a struggle between EP staged from *below* (traffic culture) and the material space (the crossroad that is planned *above*). Other traffic-related issues came up when using public transportation, for instance when tourists did not know what door to use to get into buses or trams.

Occasioned activities.

> The passenger's hobby is to take photos of flowers. Along the seaside route from Hernesaari harbour, there are several places with floral decorations, and this makes the woman excited so that she frequently stops to take pictures. (Notes taken while following an American group formed by a couple and friends, 55–60 years)

Occasioned activities (Hannam, Sheller, & Urry, 2006; Lyons & Urry, 2005) describe the situations where passengers follow own interests that become enabling elements of mobility. These situations occurred especially while walking from Länsisatama and Hernesaari harbours, where there were no classical tourist attractions, but passengers started observing everyday happenings that were somehow meaningful to them. In these situations, tourists produce new meanings and aesthetic experiences about what they find exciting while moving.

Sense of direction.

> The party arrives at Katajanokka harbour. They have problems finding the tram stop. Inside the tram, they ask a local person for some advice on how to get to the Market Square. Still, at the tram stop next to Senate Square, they go the wrong way. At the Market Square, they do not know the way back to the ship. (Notes taken while following a British group formed by a 50–55-year-old couple and the man's sister, 40–45-year-old and the mother, over 65 years old).

The process of finding a destination is different from one person to another (Golledge, 1999), and during the observation, it became evident that certain people had a good sense of direction and map reading skills, while others had frequent problems. In these situations showing features of mobile body semiotics (Jensen, 2013) the human body itself is a sort of a moving sign, and in the most positive case, a person with orientating skills operates as a wayfinding tool.

Route planning.

> The ship arrives at Länsisatama harbour, but the informant group does not know which port they are in. They start to follow the blue line that marks the walking route inside the construction site area. When the line ends, they do not know what to do, since they think that the line would have led them directly to the city centre. It takes them a long time to ponder where to go and with what mode of transportation. (Notes taken while following a group of two British couples, 55–60 years old).

In our observation, while certain passengers showed particularly careful planning, others were unprepared and did not even know the name of the port. Many passengers arriving straight into the tourist area were often just wandering around without any specific plan.

Anxiety.

> The man wants to take a bus, but other party members want to walk. He is constantly looking at his watch and making sure that others are walking briskly. (Notes taken while following a party with two couples, 55–60 years old, from Mexico).

Some examinees seemed to feel more anxiety about the time constraints than others did. These passengers were evidently more outcome-oriented tourists (Lew & McKercher, 2006) since they aimed to reach the attractions quickly. Their anxiety could be seen as an embodied performance, as they expressed worries about how much time they were spending on walking, checked the time continuously and complained when others were stopping. Anxiety was then considered as a disabling element of movement that affected the mobility of travel companions.

4. Discussion

The research has looked at different kinds of enabling and disenabling aspects of mobility that independently moving passengers faced *in situ* during their visit inside the port of call of Helsinki. Each of the three dimensions presents diverse categories related to the mobility of cruise passengers: a) PMD: wayfinding, unexpected obstacles, lack of refreshment spots, and paths' walkability; b) SI: service providers, local people, travel companions and other tourists; and c) EP: traffic culture, sense of direction, planning, time-related anxiety and occasioned activities. Overall, the results show that SI and EP elements rarely occurred on their own. Instead, they were triggered by elements of PMD.

Our study shows the evident use of wayfinding tools while passengers move inside the destination independently. Wayfinding tools are usually conceived as elements of mobility staged from *above*; however, the results indicate that *below* elements enter the scene, because installed signs can be misplaced, or unnoticed and/or unfollowed. These findings are similar to those indicated by Andriotis and Agiomirgianakis (2010) showing that cruisers missed Heraklion's attractions because of the limited information and signage in there. Also, Hamid (2014) observed in her qualitative study in Glasgow that appropriate positioning and design in street maps and signs are of upmost importance for tourism mobility.

Street signs are more meaningful in disembarking locations and in peripheries than in city centres, where tourists can use different wayfinding strategies, such as using notable landmarks that are easily traceable in paper maps. Time is always an issue for short time visitors like cruise passengers; hence, signs positioned at the harbours should indicate both distances and average walking times to tourist areas. In Jensen's (2014a) words, embodied performances and body semiotics need to be addressed in planning, so directional signs should be readable in a universal way. Having good street signs can reduce the anxiety about the time. Also, Chang (2015) in his study about wayfinding in Taipei, had identified some level of anxiety in his 86 study subjects walking through an unfamiliar city, and he also noted some individual differences, as it was evident in our observation of time-related anxiety, planning and sense of direction. A large number of informants would have wanted to spend more time in the port of call.

Our study revealed that mobile devices were used more as a wayfinding tool on public transportation. As noted again by Chang (2015), only a small group of users can see information on small screens like those of mobile phones. Using maps created various discussions and negotiations between the members of the travel parties. Hamid (2014) states that people still tend to refer to physical maps, travel books, or street maps when Wi-Fi connections are problematic and costly, especially during travels abroad. How abolishing the roaming tariffs in Europe will affect this behaviour is yet to be seen.

The results of the study show that especially while moving from the harbours located further away, there are unique features in social interaction happening amongst the travel parties, i.e. *customer to customer interaction* (Huang & Hsu, 2010). Zheng, Huang, & Li (2005) in their study on a Caribbean cruise argued that the ship is a liminal space and social relation among the passengers are a meaningful part of the cruise experience. In this current study, it was observed that while moving, travel party members whether just met on cruise or friends and family may have different interests, move with different speeds, or have different levels of orientation skills. All of these things were emphasized while the passengers were moving outside the actual tourism area. These issues are more evident when passengers have little time. For example families with children or/and elderly members had different walking speeds, and this affected the mobility of the whole group. Indeed, Nickerson and Jurowski (2001) in their survey-based study about vacationing children had similar findings about children's difficulties in following adults in cities' sightseeing.

In different types of destination, like Kuala Lumpur (Malaysia) described by Ujang and Muslim (2014), the walking experience is enriched by various street activities that provide tourists' satisfaction either psychologically and functionally. In general, pedestrians find more pleasure to walk in built environments and natural areas, when places show social and cultural vitality (Southworth, 2005). As for Helsinki, visitors find comfort in *occasioned* activities emerging from their interests and proactive roles. During our observations, passengers were trying to establish some meaningful contacts with the material environment through their embedded experiences and to create original ways of enjoying the destination. Locations allow various dynamics of interactions and performances to occur, depending on the conditions supporting their mobility experience and contributing to their enjoyment, effective time management, or anxiety. However, in general, cruise passengers' engagement with the city is heavily reduced by the time limits and the very confusion between the various locations shortly visited during the overall trip.

Cleanliness in the walkways was not considered as relevant by tourists, although we noticed that some parts of the harbours were quite untidy. About safety perception, tourists in Helsinki did not feel any disabling elements like the risk of pickpocketing, which is common in many tourist areas worldwide, or the fear of terrorist attack that is often perceived in tourist destinations (Brunt, Mawby, & Hambly, 2000) – although the Nice attack happened during the observation period.

The use of go-along observation framed within the *staging mobilities* model has supported the acquisition of different kinds of locational and qualitative data, with elements of the mobility experience interacting with others. This type of study can inform about some specific problems that are localized and having certain spatial distribution. Broader analysis on mobility and city development could support closer links between tourism planning and strategic urban planning; this goes beyond the potential of a limited technical study and enters the domain of political responsibility.

5. Conclusion

This study offers a contribution to the literature on cruise tourism and short-term mobility in ports of call. The findings suggest that the staging mobilities framework together with go-along methods allow examining enabling and disabling features of tourism mobility through more in-depth qualitative data. Although the method used was rather time-consuming and therefore, did not allow expanding the data collection further, it offers a dynamic and flexible mode of observing and classify visiting phenomena, and can be easily applied to other contexts, especially where short-term tourism is increasing. This kind of study could be also useful for urban planning in tourist cities: taking into primary consideration the spatial and functional needs of other city users and in particular, the mobility practices of the residents. Already some occasions of collision of paths, particularly along narrow pedestrian ways and bike lanes, have shown the importance of careful walkability planning. Recent international literature in both tourism and urban studies reflects on potential negative impacts that poorly planned tourism can cause in vulnerable urban areas, especially in historical centres. Like other studies have shown (Doxey, 1975; Minoia, 2017), growing tourism and particularly fast tourism can produce deeper residential issues and conflicts, even starting by occasional disturbances on specific walkways.

Acknowledgements

The authors would like to thank Charlie Hornsby, the coder of the mobile application, for his support.

Disclosure statement

No potential conflict of interest was reported by the authors.

Funding

This work was supported by The Finnish Cultural Foundation, Henrik Bitte Foundation, City of Helsinki and the Helsinki Tourism Foundation.

ORCID

Kaisa Paananen http://orcid.org/0000-0001-7710-6100

References

Ahas, R., Aasa, A., Mark, Ü., Pae, T., & Kull, A. (2007). Seasonal tourism spaces in Estonia: Case study with mobile positioning data. *Tourism Management*, *28*(3), 898–910. Retrieved from http://doi.org/10.1016/j.tourman.2006.05.010

Andriotis, K., & Agiomirgianakis, G. (2010). Cruise visitors' experience in a Mediterranean port of call. *International Journal of Tourism Research*, *12*(4), 390–404. http://doi.org/10.1002/bjs.5677/

Aranburu, I., Plaza, B., & Esteban, M. (2016). Sustainable cultural tourism in urban destinations: Does space matter? *Sustainability*, *8*(8), 1–14. Retrieved from http://doi.org/10.3390/su8080699

Baker, D. M., & Fulford, M. D. (2016). Cruise passengers' perceived value and willingness to recommend. *Tourism and Management Studies*, *12*(1), 75–85. Retrieved from http://doi.org/10.18089/tms.2016.12108

Bissell, D. (2010). Passenger mobilities: Affective atmospheres and the sociality of public transport. *Environment and Planning D: Society and Space*, *28*(2), 270–289. Retrieved from http://doi.org/10.1068/d3909

Brida, J. G., Pulina, M., Riaño, E., & Zapata-Aguirre, S. (2012). Cruise passengers' experience embarking in a Caribbean Home Port. The case study of Cartagena de Indias. *Ocean and Coastal Management*, *55*, 135–145. Retrieved from http://doi.org/10.1016/j.ocecoaman.2011.10.003

Brida, J. G., & Zapata, S. (2010). Cruise tourism: Economic, socio-cultural and environmental impacts. *International Journal of Leisure and Tourism Marketing*, *1*(2), 205–226. Retrieved from http://doi.org/10.1504/IJLTM.2010.029585

Brunt, P., Mawby, R., & Hambly, Z. (2000). Tourist victimisation and the fear of crime on holiday. *Tourism Management*, *21*(4), 417–424. Retrieved from http://doi.org/10.1016/S0261-5177(99)00084-9

Carić, H. (2016). Challenges and prospects of valuation – cruise ship pollution case. *Journal of Cleaner Production*, *111*(B), 487–498. Retrieved from http://doi.org/10.1016/j.jclepro.2015.01.033

Chang, H. H. (2015). Which one helps tourists most? Perspectives of international tourists using different navigation aids. *Tourism Geographies*, *17*(3), 350–369. Retrieved from http://doi.org/10.1080/14616688.2015.1016099

Cresswell, T. (2006). *On the Move: Mobility in the modern western world*. New York: Routledge.

Cruise Lines International Association (CLIA). (2017). Annual Report 2015. Retrieved from https://www.cruising.org/about-the-industry/research/2015-annual-report.

Cruise Market Watch. (2016). Growth. Retrieved from http://www.cruisemarketwatch.com/growth/.

De Cantis, S., Ferrante, M., Kahani, A., & Shoval, N. (2016). Cruise passengers' behavior at the destination: Investigation using GPS technology. *Tourism Management*, *52*, 133–150. Retrieved from http://doi.org/10.1016/j.tourman.2015.06.018

Doxey, G. (1975). The causation theory of visitor-resident irritants: Methodology and research inferences. In Proceedings of the Travel Research Association, 6th Annual Conference (pp. 195–198).

Diedrich, A. (2010). Cruise ship tourism in Belize. The implications of developing cruise ship tourism in an ecotourism destination. Ocean & Coastal Management, 53(5–6), 234–244

Edwards, D., & Griffin, T. (2013). Understanding tourists' spatial behaviour: GPS tracking as an aid to sustainable destination management. *Journal of Sustainable Tourism*, *21*, 580–595. Retrieved from http://doi.org/10.1080/09669582.2013.776063

Ferrante, M., De Cantis, S., & Shoval, N. (2016). A general framework for collecting and analysing the tracking data of cruise passengers at the destination. *Current Issues in Tourism*. 21(12), 1–26 Retrieved from http://doi. org/10.1080/13683500.2016.1194813

Finlay, J., & Bowman, J. (2017). Geographies on the move: A practical and theoretical approach to the mobile interview. *Professional Geographer*, *69*(4), 263–274. Retrieved from http://doi.org/10.1080/00330124.2016.1229623

Forsyth, A. (2015). What is a walkable place? The walkability debate in urban design. *URBAN DESIGN International*, *20*(4), 274–292. Retrieved from http://doi.org/10.1057/udi.2015.22

Fullagar, S., Wilson, E., Markwell, K. (2012). Starting slow: thinking through slow mobilities and experiences. In S. Fullagar, K. Markwell & E. Wilson, (Eds.), Slow tourism: Experiences and mobilities (pp. 1–11). Bristol: Channel View Publication.

G. P. Wild (International Limited and Business Research and Economic, 2015). (2015). Overview Cruise Passenger and Crew Surveys Helsinki.

Golledge, G. G. (1999). *Wayfinding Behavior: Cognitive Mapping and Other Spatial Processes*. Baltimore: JHU Press.

Hall, C. M., Le-Klähn, D.-T., & Ram, Y. (2017). *Tourism, Public Transport and Sustainable Mobility*. Bristol: Chanel View Publications.

Hall, C. M., Ram, Y. & Shoval, N. (2018). Introduction. Walking – more than pedestrian. In C. M. Hall, Y. Ram, & N. Shoval (Eds.), *The Routledge international handbook of walking*. Oxon: Routledge.

Hamid, S. A. (2014). Walking in the city of signs: Tracking pedestrians in Glasgow. *Current Urban Studies*, *2*, 263–278. Retrieved from http://doi.org/10.4236/cus.2014.23025

Hannam, K., Sheller, M., & Urry, J. (2006). Editorial: Mobilities, immobilities and moorings. *Mobilities*, *1*(1), 1–22. Retrieved from http://doi.org/10.1080/17450100500489189

Henthorne, T.L. (2000). An analysis of expenditures by cruise ship passengersin Jamaica. Journal of Travel Research, 38(3), 246–250

Huang, J., & Hsu, C. H. C. (2010). The impact of customer-to-customer interaction on cruise experience and vacation satisfaction. *Journal of Travel Research*, *49*(1), 79–92. Retrieved from http://doi.org/10.1177/0047287509336466

Jaakson, R. (2004). Beyond the Tourist Bubble? Cruiseship Passengers in Port. *Annals of Tourism Research*, *31*(1), 44–60. Retrieved from http://doi.org/10.1016/j.annals.2003.08.003

Jensen, O. B. (2013). *Staging Mobilities*. Oxon: Routledge.

Jensen, O. B. (2014a). *Designing Mobilities*. Aalborg: Aalborg Universitetsforlag.

Ujang & Muslim (2014). Walkability and attachment to tourism places in the City of Kuala Lumpur, Malaysia. In ATINER's conference paper series, March 2014 (pp. 1–17).

Jensen, O. B. (2014b). *Mobilities design – towards a "new materials turn."* Mobilities and design Work Shop, Lancaster, UK.

Jordan, E. J., & Vogt, C. A. (2017). Residents' perceptions of stress related to cruise tourism development. *Tourism Planning and Development, 14*(4), 527–547. Retrieved from http://doi.org/10.1080/21568316.2017.1287123

Judd, D. R. (1999). Constructing the Tourist Bubble. In S. Judd & D. Feinstein (Eds.), *The Tourist City* (pp. 35–53). New Haven: Yale University Press.

Kang, S.(2016). Associations between space-time constrains and spatial patterns of travels. Annals of Tourism Research, 61, 127–141.

Kusenbach, M. (2003). Street phenomenology. The Go-Along as ethnographic research tool. *Ethnography, 4*(3), 455–485. Retrieved from http://doi.org/10.1177/146613810343007

Kusenbach, M. (2012). Go-Alongs. In S. Delamont (Ed.), Handbook of qualitative research in education (pp. 252–264). Cheltenham: Edward Elgar Publishing

Lanng, D. B., & Jensen, O. (2016). Linking wayfinding and wayfaring. In R.H. Hunter, L.A. Anderson & B. Belza (Eds.), Community wayfinding: Pathways to understanding (pp. 247–260). Switzerland: Springer International Publishing

Larsen, J. (2001). Tourism mobilities and the travel glance: Experiences of being on the move. *Scandinavian Journal of Hospitality and Tourism, 1*(2), 80–98. http://doi.org/10.1080/150222501317244010

Larsen, J., Axhausen, K. W., & Urry, J. (2006). Geographies of social networks: Meetings, travel and communications. *Mobilities, 1*(2), 261–283. Retrieved from http://doi.org/10.1080/17450100600726654

Larsen, S., & Wolff, K. (2016). Exploring assumptions about cruise tourists' visits to ports. *Tourism Management Perspectives* 17, 44–49. Retrieved from http://doi.org/10.1016/j.tmp.2015.12.001

Lew, A., & McKercher, B. (2006). Modeling tourist movements: A local destination analysis. *Annals of Tourism Research, 33*(2), 403–423. Retrieved from http://doi.org/10.1016/j.annals.2005.12.002

Lynch, K. (1960). *The Image of the City*. Cambridge(MA): The MIT Press.

Lyons, G. & Urry, J. (2005). Travel time use in the information age. Transportation Research A, 39(2), 257–276.

Manning, T. (2006). Managing cruise ship impacts: Guidelines for current and potential destination communities a backgrounder for prospective destination communities. Retrieved from http://www.tourisk.org/content/projects/Managing Cruise Ship Impacts.pdf

Marti, B. E. (1990). Geography and the cruise ship port selection process. *Maritime Policy and Management, 17*(3), 157–164. Retrieved from http://doi.org/10.1080/03088839000000023

McCalla, R. (1998). An investigation into site and situation: Cruise ship ports. Tijdschrift voor Economische en Sociale Geografie, 89(1), 44–55.

McKercher, B., Shoval, N., Ng, E., & Birenboim, A. (2012). First and repeat visitor behaviour: GPS tracking and GIS analysis in Hong Kong. *Tourism Geographies, 14*(1), 147–161. Retrieved from http://doi.org/10.1080/14616688.2011.598542

Minoia, P.(2017). Venice reshaped? Tourist gentrification and sense of place. In N. Bellini & C. Pasquinelli (Eds.), Tourism in the City. Towards an integrative agenda on urban tourism (pp. 261–274). Cham: Springer.

Nickerson, N. P., & Jurowski, C. (2001). The influence of children on vacation travel patterns. *Journal of Vacation Marketing, 7*(1), 19–30. Retrieved from http://doi.org/10.1177/135676670100700102

Papathanassis, A., & Beckmann, I. (2011). Assessing the "Poverty of Cruise Theory" hypothesis. *Annals of Tourism Research, 38*(1), 153–174. Retrieved from http://doi.org/10.1016/j.annals.2010.07.015

Pearce, P.L. (2005). Tourist behaviour: Themes and conceptual schemes. Clevedon: Chanel View Publications.

Penco, L., & Di Vaio, A.. (2014). Monetary and non-monetary value creation in cruise port destinations: an empirical assessment. *Maritime Policy and Management, 41*(5), 501–513. Retrieved from http://doi.org/10.1080/03088839.2014.930934

Port of Helsinki (2017). Annual Report 2016. Retrieved from http://www.portofhelsinki.fi/sites/default/files/attachments/Vuosikertomus 2016 englanti low res.pdf

Ram, Y., & Hall, C. M. (2018). Walkable places for visitors. Assessing and designing for walkability. In C. M. Hall, Y. Ram, & N. Shoval (Eds.), *The Routledge international handbook of walking*. Oxon: Routledge.

Rodrigue, J. P., & Notteboom, T. (2013). The geography of cruises: Itineraries, not destinations. *Applied Geography, 38*(1), 31–42. Retrieved from http://doi.org/10.1016/j.apgeog.2012.11.011

Samarasekara, G. N., Fukahori, K., & Kubota, Y. (2011). Environmental correlates that provide walkability cues for tourists: An analysis based on walking decision narrations. *Environment and Behavior, 43*(4), 501–524. Retrieved from http://doi.org/10.1177/0013916510379350

Satta, G., Parola, F., Penco, L., & Persico, L. (2015). Word of mouth and satisfaction in cruise port destinations. *Tourism Geographies, 17*(1), 54–75. Retrieved from http://doi.org/10.1080/14616688.2014.938689

Sheller, M., & Urry, J. (2006). The new mobilities paradigm. *Environment and Planning A, 38*(2), 207–226. Retrieved from http://doi.org/10.1068/a37268

Shoval, N., & Ahas, R. (2016). The use of tracking technologies in tourism research: The first decade. *Tourism Geographies, 18*(5), 587–606. Retrieved from http://doi.org/10.1080/14616688.2016.1214977

Shoval, N., & Isaacson, M. (2007). Tracking tourists in the digital age. *Annals of Tourism Research, 34*(1), 141–159. Retrieved from http://doi.org/10.1016/j.annals.2006.07.007

Silvestre, A. L., Santos, C. M. & Ramalho, C. (2008). Satisfaction and behavioural intentions of cruise passengers visiting the Azores. *Tourism Economies, 14*(1), 169–184. Retrieved from http://doi.org/https://doi.org/10.5367/000000008783554802

Southworth, M. (2005). Designing the walkable city. *Journal of Urban Planning and Development, 131*(4), 246–257. Retrieved from http://doi.org/10.1061/(ASCE)07339488(2005)131:4(246)

Sönmez, S. F., & Graefe, A. R. (1998). Influence of terrorism risk on foreign tourism decisions. *Annals of Tourism Research, 25*(1), 112–144. Retrieved from http://doi.org/10.1016/S0160-7383(97)00072-8

Teye, V. B., & Leclerc, D. (1998). Product and service delivery satisfaction among North American cruise passengers. *Tourism Management, 19*(2), 153–160. Retrieved from http://doi.org/10.1016/S0261-5177(97)00107-6

Tranberg Jensen, M., Gyimothy, S., & Jensen, O. B. (2015). Staging interrail mobilities.. *Tourist Studies, 16*(2), 111–132. Retrieved from http://doi.org/10.1177/1468797615594740

Ujang & Muslim (2014). Walkability and attachment to tourism places in the City of Kuala Lumpur, Malaysia. In ATINER's conference paper series, March 2014 (pp. 1–17).

Urry, J. (2000). Mobile Sociology 1. *The British Journal of Sociology, 51*(1), 185–203. Retrieved from http://doi.org/10.1111/j.1468-4446.2000.00185.x

Urry, J. (2010). Mobile *Sociology. British Journal of Sociology, 61*(s1), 347–366. Retrieved from http://doi.org/10.1111/j.1468-4446.2009.01249.x

Veijola, S., & Falin, P. (2016). Mobile neighbouring. *Mobilities, 11*(3), 382–399. Retrieved from http://doi.org/10.1080/17450101.2014.936715

Visit Helsinki. (2017). Helsingissä kävi kesällä 409 000 kansainvälistä risteilymatkustajaa (Helsinki was visited by 409 000 international cruise ship passengers in the year 2016). Retrieved from http://www.visithelsinki.fi/fi/ammattilainen/media/uutishuone/helsingissa-kavikesalla-409-000-kansainvalista-risteilymatkustajaa

Wood, R. E. (2000). Caribbean cruise tourism: Globalization at sea. *Annals of Tourism Research, 27*(2), 345–370. Retrieved from http://doi.org/10.1016/S01607383(99)00073-0

Xie, H. J., Kerstetter, D. L., & Mattila, A. S. S. (2012). The attributes of a cruise ship that influence the decision making of cruisers and potential cruisers. *International Journal of Hospitality Management, 31*(1), 152–159. Retrieved from http://doi.org/10.1016/j.ijhm.2011.03.007

Yarnal, C. M., & Kerstetter, D. (2005). Casting off: an exploration of cruise ship space group tour behavior, and social interaction. *Journal of Travel Research*, *43*(4), 368–379. Retrieved from http://doi.org/10.1177/0047287505274650

Zheng, W., Huang, X., & Li, Y. (2017). Understanding the tourist mobility using GPS: Where is the next place? *Tourism Management*, *59*, 267–280. Retrieved from http://doi.org/10.1016/j.tourman.2016.08.009

OPEN ACCESS

Airbnb as an instigator of 'tourism bubble' expansion in Utrecht's Lombok neighbourhood

Dimitri Ioannides, Michael Röslmaier and Egbert van der Zee

ABSTRACT

The Airbnb phenomenon as part of the broader growth of the so-called collaborative economy has grabbed the attention of a growing number of tourism researchers. Among the topics explored have been investigations as to the spatial tendencies of Airbnb in cities and discussions concerning its effects, *inter alia*, on gentrification, over-touristification and eventual resident displacement. Recognizing that the majority of extant studies have been conducted either in major cities, which in their own right attract large numbers of visitors or in tourism-intensive smaller communities we chose to investigate what Airbnb growth means for a mid-sized city with a highly diversified economy, which is not yet over-touristified. Our focus was on the Dutch city of Utrecht. Through a geospatial and statistical analysis of AirDNA data, we explored the growth of Airbnbs in the city overall, focusing specifically on the phenomenon's effects on the Lombok neighbourhood, a nascent 'neo-bohemia' neighbouring the city-centre tourist bubble. Our analysis reveals that although Airbnb activity in this neighbourhood is relatively recent there are signs suggesting that further touristification of parts of Lombok has ignited increased Airbnb activity. Moreover, there is a distance decay of Airbnb activity as one moves away from the city centre and from established tourism services including restaurants. These findings suggest that in an emerging neo-bohemian space such as Lombok, Airbnb takes on a role as instigator of urban tourism bubble expansion. The study ends with a call for further investigations to better understand the implications expanded Airbnb activity has, among others, on social justice within cities. For example, future investigations could examine the manner in which Airbnbs influence the everyday life of the residents of urban spaces and investigate the conflicts that might arise in Airbnb ghettoes between visitors and locals.

Introduction

Airbnb has emerged in recent years as a global *tour de force* in the tourist fabric of numerous places (Dredge & Gyimothy, 2017; Oskam & Boswijk, 2016; Sundararajan, 2015). This and similar innovations in the hospitality sector such as HomeAway, 9Flats or Housetrip constitute online peer-to-peer platforms enabling individuals to easily convert their properties (i.e. either entire homes or just one or two rooms) into short-term rentals. The advantage to the owners of these properties is that through the online platform they can easily reach a global market. Simultaneously, by using Airbnb, visitors have access to an ever-increasing range of accommodation options while travelling (Guttentag, 2017).

The proliferation of Airbnb and accommodations like it has transformed entire city blocks into areas catering to short-term visitors, frequently making it harder for locals to find affordable permanent housing (Füller & Michel, 2014; Colomb & Novy, 2017; Oppilard, 2017). Lee (2016) reinforces this point, arguing that landlords find it more lucrative to rent out their properties to visitors rather than local inhabitants or students. This can lead to severe housing shortages, which is a possible factor behind skyrocketing rents (Arias Sans & Quaglieri Dominguez, 2016; Lee, 2016). Others, however, (e.g. Füller & Michel, 2014) argue that increased housing prices in particular neighbourhoods can only partially be explained by the drastic expansion of short-term accommodation and that the complete explanation is far more complex.

The thorny question is what do these developments mean for the future of metropolitan areas and their populations worldwide? After all, Airbnb and similar companies face growing resentment of locals who fear that these developments, together with many other tourist-related activities, transform their once-tranquil residential neighbourhoods into visitor ghettoes (Colomb & Novy, 2017). Meanwhile, municipalities scramble to identify ways to regulate Airbnb's growth either through taxation or by imposing drastic measures aimed at limiting or entirely eradicating short-term rentals (Oppilard, 2017). For instance, Airbnb's phenomenal expansion in Reykjavik, has led the Icelandic government to impose curbs on transforming more homes and rooms into short-term rentals (Baranjuk, 2017). Similarly, Berlin and Amsterdam limit how long properties can be rented out through Airbnb (Comiteau, 2016; Füller & Michel, 2014).

To further understand how Airbnb influences host communities, it is important to study the pattern of its spatial expansion throughout the urban environment. Where does

Airbnb activity proliferate? Do Airbnbs concentrate within the so-called downtown 'tourism bubbles' (Edensor, 1998; Judd, 2003), which are 'areas insulated from the larger urban milieu' (Füller & Michel, 2014, p. 1305)? Alternatively, do they spread to other parts of the city? If so, what characteristics do neighbourhoods with Airbnbs have?

Several studies have explored these topics to varying degrees. For example, Arias Sans and Quaglieri Dominguez (2016) have shown how Airbnb activity in Barcelona overlaps with heavily touristified areas. Similarly, Gutierrez, Garcia-Palomares, Romanillos, and Salas-Olmedo (2017) have conducted a detailed investigation of the spatial distribution of Airbnb versus hotels in the same city. Yrigoy (2016) demonstrates that Airbnb activity in Palma de Mallorca concentrates overwhelmingly on the central city and in the case of New York City, Dudás, Vida, Kovalcsik, and Boros (2017) have noted the clustering of Airbnbs in neighbourhoods where the population is young, there are many homes and close proximity to attractions.

The common thread linking the aforementioned studies is either their focus on large cities with considerable tourist activity or smaller (often historical) cities, where tourism dominates the economy. Our own point of departure was to identify how Airbnb activity plays out in a mid-sized city, which is not immediately associated with heavy tourist flows. In other words, how does Airbnb activity evolve in places where tourism is only part of a diversified economy? To do this, we conducted a case study of Utrecht in the Netherlands, which is perhaps most famous for its university and being a major hub of the Dutch railway system. This is a historic city with a wealth of attractions and yet, according to Lonely Planet (n.d.) it is 'bizarrely under-visited'. We were especially curious to see how Airbnb expands outward from the historic city centre into the surrounding neighbourhoods, by focusing on the part of the city known as Lombok. Lying close to the city centre, this ethnically mixed community has functioned for several years as a retail hub for the city's diverse population. There are signs, however, that part of Lombok is becoming touristified with the emergence of upscale restaurants and shopping establishments. We examine how Airbnb is infiltrating the area and contemplate whether this phenomenon cements the expansion of the downtown tourism bubble within Lombok.

The paper is structured as follows. Following a review of the pertinent literature regarding the rise of the collaborative economy and the significance of short-term accommodation platforms like Airbnb we focus on our case study. We explain why Utrecht was selected as a case study and how we gathered the relevant data. The analysis of our findings is followed by a discussion and concluding remarks.

The rise of the collaborative economy and its implications for tourism

Dredge and Gyimóthy (2017) argue that contemporary tourism is increasingly influenced by the so-called collaborative economy, which according to the European Commission (2016, p. 3)

> refers to business models where activities are facilitated by collaborative platforms that create an open marketplace for the temporary usage of goods and services often provided by private individuals ... [It] involves three categories of actors: (service providers who share assets, resources, time and/or skills – these can be private individuals offering services on an occasional basis ('peers') or service providers acting in their professional capacity ('professional service providers'); (ii) users of these; and (iii) intermediaries that connect – via an online

> platform – providers with users and that facilitate transactions between them ('collaborative platforms'). Collaborative economy transactions generally do not involve a change of ownership and can be carried out for profit or not-for-profit.

While additional terms are used to discuss this phenomenon, including 'sharing economy', 'gig economy', 'access economy' and 'on demand economy' (Dredge & Gyimóthy, 2017), to avoid confusion we talk exclusively about the 'collaborative economy'. The aforementioned European Commission report indicates that although this arguably remains a small part of the overall economy, sectors such as short-term accommodation rentals, passenger transport, household services, professional and technical services, and collaborative finance demonstrate substantial growth. Although it could be argued that transactions fitting the concept of 'sharing' such as the exchange of goods (bartering) have existed historically – and continue to thrive in some societies – (Belk, 2010), the fundamental force leading to the current acceleration of the collaborative economy has undoubtedly been the phenomenal diffusion of digital technologies. Effectively, anyone with a smart phone or a computer who puts her mind to it has the potential to become an entrepreneur offering goods and services to a global market. These can be provided for a fee but also include transactions where the entrepreneurs offers something in exchange for something else.

In this paper, we specifically focus on the collaborative economy's implications for tourism, a matter that has only recently grabbed the attention of scholars (Dredge & Gyimóthy, 2017; Forno & Garibaldi, 2015; Guttentag, 2015; Molz, 2013; Sigala, 2015; Tussyadiah & Pesonen, 2015). Tourism itself possesses several inherent characteristics, which drive many stakeholders within this sector to embrace the collaborative economy. Dredge and Gyimóthy (2015, p. 296) explain that 'redundancy is present in existing tourism systems in the form of dead capital, idling assets and latent expertise' meaning that 'local expertise and knowledge can be monetized by offering guided tours or dining experiences with local hosts'. In other words, 'technology-mediated platforms', enable the owners of assets such accommodation or private vehicles to respectively 'share' these for a fee.

Short-term accommodation rentals: the case of Airbnb

A key player in the collaborative economy is Airbnb, which Guttentag (2017, p. 97) labels a 'disruptive innovation' whose expansion in popularity signifies a transformation 'from a niche product into a mainstream one' (p. 98). Since it was established in 2008, Airbnb has grown rapidly. Today, approximately 4 million listings are found worldwide, a figure which astonishingly surpasses the 3.9 million rooms accounted for by the five largest hotel chains (Wood, 2017).

Individuals list their properties through Airbnb, aiming to utilize existing resources (i.e. their home) to expand their household income with limited investment. In this manner, the online platform encourages small-scale entrepreneurship (Gyimóthy, 2017). From the tourist's perspective, Airbnb opens up a diverse range of affordable accommodation alternatives. In Amsterdam, it provides cheaper stays than hotels (Kathan, Matzler, & Veider, 2016; see also Guttentag, 2015) Saving money through Airbnb encourages budget-minded travellers to overnight in upscale areas where hotels are expensive but also to stay longer or travel more frequently (Tussyadiah & Pesonen, 2015). Furthermore, these visitors might engage in activities they normally would have avoided had they chosen more expensive accommodation. Finally, Guttentag (2015, 2016, 2017) regards Airbnb's appeal to its users as representative of MacCannell's (1999) point that travellers purposefully seek out local and authentic spaces. In fact, Airbnb itself argues that holidaying in real homes allows one to become more like a local.

Debates concerning short-term accommodation rental platforms

Despite the perceived benefits associated with Airbnb, several writers criticize this and platforms like it. Specifically, they suggest that Airbnb negatively affects the housing market, hotel industry, hospitality workers, neighbourhoods and revenues in various communities within, *inter alia*, the United States, Netherlands, Spain, Sweden and Finland (Fang, Ye, & Law, 2016; Gant, 2015; Interian, 2016; Lee, 2016; Lehr, 2015; Neeser, 2015; Zervas & Byers, 2016). Zervas and Byers, for example, believe it leads to declining occupancy rates in existing hotels (especially budget and mid-range establishments) and this, consequently, reduces these establishments' daily intake.

Others regard Airbnb as a gentrifying force in cities like Barcelona, New Orleans and San Francisco, arguing that it leads to rising rents and seriously reduces access to affordable housing (Fang et al., 2016; Gant, 2015; Johnson, 2015). Lee (2016) stresses that Los Angeles' affordable housing crisis is worsened because of short-term rentals. He believes that Airbnb creates a situation whereby millions of tourists compete with potential long-term renters for the identical housing stock. Lee states 'each apartment or home listed year-round on Airbnb is a home that has been removed from the residential housing market and added to the city's aggregate stock of hotel rooms' (p. 234). His point is that when property owners realize they can maximize their profits from Airbnb they are less likely to make their homes available for long-term leases. Meanwhile, Johnson (2015) attributes New Orleans' impressive growth in Airbnbs to (1) homeowners' tendency to use the platform to enhance their incomes and (2) importantly, the involvement of property companies in the short-term rental market. He maintains that the result is that 'the rapid spread of Airbnb rentals has altered the character of neighbourhoods' (p. 193).

Füller and Michel (2014) cautiously question the short-term holiday rental sector's role in neighbourhood gentrification. Referring specifically to Berlin's Kreuzberg district, they attribute rising housing costs to the increase in households wishing to own homes in the area where there is a fixed supply of residences and the fact that rental properties are being transformed into owner-occupied units, many of which are becoming second homes. Thus, they suggest that the growth in the area's holiday rentals is a consequence of the aforementioned forces.

Regardless as to whether the second home rental sector is a dominant reason of gentrification or merely a result of a far more complex set of broader forces, to further understand Airbnb's effects amid broader processes of touristification, it is worth revisiting the broader urban tourism literature (see Füller & Michel, 2014; Gotham, 2005; Hoffman, Fainstein, & Judd, 2003). In the following paragraphs, we briefly revisit issues and debates on this topic, specifically focusing on the hypothesized shift from neoliberal top-down-driven, major-scale and standardized downtown tourism projects to the appearance of new forms of urban tourism occurring in regular residential areas (Ioannides & Petridou, 2016;

Maitland, 2007). We ask whether or not activities such as Airbnb spread to these new zones of urban tourism and, if so, explore the implications for these neighbourhoods?

From tourism bubble to neo-bohemia

Much has been written over the last two decades concerning tourism's growth as a key redevelopment strategy for cities in the throes of economic restructuring (Fainstein, Hoffman, & Judd, 2003). Referring specifically to the United States, Hackworth (2007) argues that since the 1970s the so-called 'neoliberal spatial fix' (p. 79) has focused heavily on the revival of downtown areas through major real-estate undertakings motivated by large incentive packages. Thus, traditional downtowns and surrounding districts once associated with production (e.g. old factories, and waterfront warehouse areas) are transformed into spaces of consumption represented through shopping centres, luxury housing, and visitor attractions (Ioannides & Petridou, 2016). These redevelopments, variations of which appear in cities throughout the world are dubbed 'tourist bubbles' or 'enclaves' (Edensor, 1998; Judd, 2003) presenting the characteristics of highly regulated, predictable and sterile spaces, dominated by symbols of multinational capital (e.g. chain establishments) and displaying little – if any – connection to the rest of the city (Edensor, 1998; Fainstein et al., 2003; Füller & Michel, 2014). Although, to be sure, such places are not necessarily physically walled-off from other parts of the city, in the manner of a gated community or an all-inclusive tourism enclave, we can perceive of them as tourist bubbles, precisely because of their obvious disconnect to the rest of the city. The high level of standardization and regulation marking these spaces, often reflecting well-recognized images of global capital, signifies that they bear minimal association to the communities within which they are found (Ioannides & Petridou, 2016).

Yet, clearly many people choose to escape these tourist bubbles. Judd (2003) notes that they do so, precisely because they seek someplace more adventurous, places he describes as 'edgy' (p. 30), where encounters and experiences are not scripted as they can be within the bubble. These are once off-the-beaten-track parts of the city where locals go about their own regular lives. It is precisely these types of 'mundane activities' (Maitland, 2007, p. 176) that entice the visitors rather than any deliberate tourism-oriented attraction. Indeed, within these sought-after spaces the actions both visitors and locals engage in cannot be readily distinguished (Füller & Michel, 2014; Judd, 2003) since both groups belong to the 'cosmopolitan consuming class' (Fainstein et al., 2003, p. 243; see also Maitland, 2007).

Lloyd (2006) traces Chicago's Wicker Park's gradual transformation from a transitional area offering a mix of warehouses and low-income minority housing into an alternative neo-bohemian district, a few metro stops to the northwest of the central business district. Here, one finds entertainment venues and non-chain eating establishments, cafes and absinthe bars, which lure the sophisticated residents and tourists who are escaping the stereotypical visitor bubbles. Ioannides and Petridou (2016, p. 29) attribute the attraction of Wicker Park and similar neo-bohemias to their transmission of 'a feeling of far greater originality ... precisely because they retain much of their original (albeit refurbished) urban fabric and, importantly, because the visitor mingles with the locals representing various ages, socioeconomic levels, and ethnicities'.

Airbnb as a means to expanding the bubble?

If neo-bohemias entice visitors wishing to avoid the city-centre tourist bubbles, it stands to reason that they would also be places where there is an emergence of short-term rentals, including Airbnb properties, especially when there is a distinct shortage or absence of formal (commercial) tourism accommodation (i.e. hotels). Precisely, because the Airbnb phenomenon is one that responds rapidly to growing demand, it can be an important force of further touristification in such neighbourhoods. Indeed, one could argue that Airbnb is a multinational force – albeit one camouflaged by the absence of physical signs as to its existence – which reinforces the expansion of central city tourist bubbles into surrounding areas precisely because it eventually leads to commercialized areas with few long-term residents.

The Airbnb company itself (referred to in Arias Sans & Quaglieri Domìnguez, 2016) boasts that in Barcelona, Airbnb properties are likelier to locate outside neighbourhoods with hotels. This allows visitors to experience more than the stereotypical tourist sights on offer. Our cursory examination of the website Inside Airbnb (2017) reveals noticeable hubs of Airbnb activity in Chicago's Wicker Park, Melbourne's Fitzroy district, London's Islington and Barcelona's Vila de Gràcia all of which match the neo-bohemian profile. Similarly, Füller and Michel (2014) have pointed out that one of Berlin's 'new urban tourism' (p. 1309) areas, Kreuzberg, has seen a noticeable increase in short-term holiday flats stressing that such accommodations, rather than formal vacation rentals (e.g. hotels and hostels) give the visitor the sense 'of being seamlessly embedded in a local urban neighbourhood' (p. 1311).

These examples do not reflect, of course, that Airbnb spreads uniformly throughout the city. Rather, Airbnb activity is observed only in particular areas, which are either close to existing hubs of tourism activity or within districts with considerable visitor-related activity already (e.g. areas that have a reputation for night life). Referring specifically to Barcelona, Arias Sans and Quaglieri Domìnguez (2016) indicate a notable Airbnb cluster in and around the city centre itself. By contrast, several neighbourhoods throughout the city have little or no activity thus disputing Airbnb's assertions. Indeed, the dominant hub is within Ciutat Vella (the old city), coinciding precisely with the highest concentration of hotels. In other words, Airbnb operators cluster around existing holiday accommodations and *de facto* within or very close to well-established visitor attraction zones. A secondary cluster is found in the el Raval neighbourhood but Arias Sans and Quaglieri Domìnguez attribute this to the high proximity of the popular touristified zone of Las Ramblas. Interestingly, the Airbnb property descriptions for el Raval praise the city's main 'hotpots' (Arias Sans & Quaglieri Domìnguez, 2016, p. 217) and provide very little publicity about the immediate neighbourhood.

Only in La Vila de Gràcia with its secondary cluster of Airbnb activity do most property descriptions highlight the neighbourhood's charms. Explaining this as the exception rather than the norm, Arias Sans and Quaglieri Domìnguez (2016) argue that this well-established neo-bohemia is already a *bona fide* touristified zone and a popular side-destination for many visitors to Barcelona. Overall, their findings debunk the Airbnb corporation's argument that most of their properties are located outside touristic zones allowing the users to experience life in regular neighbourhoods. Indeed, the correlation between the presence of hotels and the presence of Airbnb properties in Barcelona is high. They also dispute the assertion that users of Airbnb activities get to live like locals since the areas where these properties are located attract an international population including large groups of expatriates.

The preceding paragraphs illustrate our limited understanding of Airbnb's geographic spread throughout a community. Obviously, some parts of a city are noticeable Airbnb clusters as opposed to others and, certainly, historic centres are magnets for such activity. Neo-bohemian neighbourhoods seem to emerge as Airbnb hotzones though these properties are not uniformly spread throughout these areas. Questions to be investigated further include: (1) what effect if any does the distance from existing tourist bubbles have on Airbnb activity (i.e. is there a distance-decay effect); (2) to what extent does proximity to tourist infrastructure such as hotels and restaurants influence whether or not there is Airbnb activity?

In the next section, we investigate these issues through an examination of Airbnb activity in the Dutch city of Utrecht. We specifically focus on Lombok, a neighbourhood fitting the characteristics of an emerging neo-bohemia. Our aim is to investigate how Airbnb activity plays out in the city overall and how it influences Lombok, as well as to unravel the forces that are influencing the emergent patterns.

Study area selection and method

Utrecht

Our case study is Utrecht and, specifically, the Lombok neighbourhood located to the west of the city centre and separated from it by railway tracks. We purposefully selected this mid-size city rather than a larger metropolitan region (e.g. Barcelona and San Francisco), which often constitute the focus of investigations regarding Airbnb activity. Also, we avoided heavily touristified historic tourist towns such as Oxford, Bath or Bruges, which are completely overrun by visitors by choosing a city where tourism growth, though noticeable in recent years, is still at a relatively early stage despite the attractions on offer.

Utrecht, a major university town has a historic centre featuring several attractions including a medieval canal, churches and the university's ceremonial buildings (Lonely Planet, n.d.). There are several popular museums, a lively bar and restaurant scene, a 100,000 m^2 trade fair facility and the busiest transportation hub in the Netherlands (the central station). The city boasts a UNESCO world heritage site (the Rietveld Schröder residence) just to the east of the centre. Each year, approximately 4 million international and Dutch visitors are drawn to the city (the vast majority are day-visitors, presumably because of Utrecht's proximity to Amsterdam and Rotterdam), while 540,000 registered overnight visitors were recorded in 2015. Most overnight visitors are business travellers (approximately 300,000) (Toerisme Utrecht, 2015). Meanwhile, the stock of available hotel rooms is surprisingly low considering the level of visitation (only 1749 in 2014) and struggles to cope with peak-pressure during events, conferences and trade fairs (Horwath, 2014). In 2014, the average annual occupancy rate was 72% while the mean room price was €93, rating Utrecht second only after Amsterdam in the ranking of the accommodation sector in major Dutch cities (Horwath, 2016). By contrast, the informal (peer-to-peer) accommodation sector is sizeable and absorbs the peak pressure for accommodation in the city during major events (e.g. the Tour de France' Grand Depart in July 2015). Airbnb is included within this informal peer-to-peer accommodation sector, which also accounts for other similar platforms and couch-surfing.

Within Utrecht itself lies Lombok, a densely populated, former working-class neighbourhood built in the late nineteen and early twentieth century to accommodate factory workers. The area was set for renewal in the 1970s but because of an influx of squatters and many migrants, especially from Morocco and Turkey, it retained its working class residential feel. Much of the housing stock was seriously dilapidated, with several units lacking proper sanitation, heating, and hot water. From the 1990s onwards, Lombok witnessed waves of gentrification, precisely due to its central location, aesthetic and architectural features. The rise in the neighbourhood's popularity, especially among highly educated households had much to do with its cosmopolitan vibe (Permentier, Van Ham, & Bolt, 2008).

Lombok displays characteristics demonstrating its slow but gradual transformation into a neo-bohemia, certainly one catering to new urban tourism (Füller & Michel, 2014). It entices both residents from other parts of the city and out-of-town visitors. Visit Utrecht explains that the neighbourhood's proximity to the Central Station and the city centre adds to the area's attraction. While its main shopping street, Kanaalstraat, retains a multi-ethnic feel with several shops and other businesses catering to ethnic minorities and students, there has been a recent shift whereby a growing number of establishments serves an upscale cosmopolitan clientele. At the neighbourhood's western edge, closest to the city centre, several trendy restaurants have opened in buildings such as refurbished factories. Examples include fashionable street-food cafes and world cuisine and fusion restaurants. Also, there are cultural initiatives such as a museum café (exhibiting the local history) and an old sawmill celebrating the area's industrial heritage. No major hotels are present in the neighbourhood but there is a single B&B establishment. Two chain-hotels lie just beyond Lombok's border with the city centre. Conversely, the offer of peer-to-peer holiday accommodation, though limited, is growing.

Data selection

Because Airbnb data are proprietary, we utilized information purchased through the intermediary service, AirDNA, which obtains data on Airbnb presence and usage through scraping the Airbnb website daily. Using algorithms, AirDNA estimates the following: whether or not Airbnb accommodations are rented out; for how long; at what price and; whether the accommodations are blocked off by the owner? AirDNA's advantage is that the daily scraping leads to an estimate of Airbnb activity, while other services such as Inside Airbnb gather data infrequently, only revealing numbers of accommodations.

The data-set used contained all active properties on Airbnb since the data scraping began (20 August 2015). It contains a wealth of information about: the host; the property itself; the listing; and activity. We referred to two data-sets, one with information collected during a full year since the first day of scraping (until 20 August 2016) (AirDNA, 2016) and one ending on 22 October 2017 (AirDNA, 2017). Using more than one data-set enables a comparison for Utrecht of the evolution of Airbnb in the period August 2015 to October 2017.

Airbnb accommodations identified as part of the 'formal' visitor accommodation sector were filtered out. Thus, we subtracted hotels and other registered visitor accommodation

Table 1. Utrecht's Airbnb profile focusing on Lombok and the Historic city centre (August 2016 and October 2017).

Neighbourhood characteristics (2016)	Utrecht (entire city)		Historic city centre		Lombok	
Surface area	99.21 km^2		1.42 km^2		1.49 km^2	
Inhabitants	338,967		14,370		19,085	
Number of households	174,765		9950		11,280	
Percentage non-western migrants	22.0%		8.5%		21.9%	
Airbnb characteristics	**August 2016**	**October 2017**	**August 2016**	**October 2017**	**August 2016**	**October 2017**
Number of active Airbnbs	1202	2156	221	415	152	224
Active Airbnbs per 100 households	0.69	1.23	2.22	4.17	1.35	1.96
Number and percentage of very active Airbnbs (>60 days rented out)	199 (16.6%)	832 (38.6%)	40 (18.2%)	165 (39.8%)	22 (14.5%)	88 (39.3%)
Number and percentage of Professional Airbnbs (Private home AND owned by host offering multiple listings AND available for more than 60 days)	94 (7.8%)	214 (9.9%)	26 (11.2%)	53 (12.8%)	14 (9.2%)	27 (12.1%)
Average realized price per night	€98	€97	€125	€115	€90	€97
Average amount of bookings (nights) per active Airbnb	7.4 (34.2)	16.4 (66.7)	8.4 (34.2)	20.6 (74.0)	6.1 (30.6)	16.6 (69.4)
Average yearly income per active Airbnb	€3100	€6300	€4000	€8000	€2500	€6400

Sources: Central Bureau for Statistics (2017); AirDNA (2016, 2017).

that list their rooms on the Airbnb platform, instead focusing on the 'informal' Airbnb sector, meaning private homes or rooms within private homes listed on the online platform. 'Active' properties are those rented out at least once in the 12 months prior to the data collection and 'very active Airbnbs' the ones rented out for more than 60 days per year. 'Professional Airbnbs' refers to entire homes/apartments, controlled by owners with multiple listings (two or more), available for over 60 days per year (see Table 1).

For reliability, we cross-referenced our data-set to a sample of 30 Airbnb accommodations that shared information with us. This revealed that the Utrecht data-set was accurate on the listing and accommodation information (all accommodations were present, and all information was scraped correctly). However, the Airbnb locations varied slightly compared to the precise addresses, as Airbnb distorts these to safeguard user privacy. Also, the AirDNA data-set was less accurate on accommodation activity variables since, on average, there was an overestimation on booking numbers and number of nights and yearly income. Airbnb accommodations with high cancelation rates or those frequently blocked out for short periods proved less reliable, probably due to the inability of the scraping algorithm to distinguish between actual bookings and cancellations and/or (short) blocked-off periods. This means that the actual average amount of bookings, nights and incomes should be interpreted cautiously since, in reality, they are likely to be between 8% and 15% lower. The data were mapped and analysed using ArcGIS 10.4.

Additionally, we applied a multivariate regression analysis to uncover the factors influencing the distribution of Airbnb over the city. The Dutch Central Bureau for Statistics (CBS, 2017) offers a rich data-set of statistics on sub-neighbourhood level. These statistics give an insight in, among other characteristics, demographic composition, socio-economic status and vicinity of various services for these sub-neighbourhoods. In total, Utrecht consists of 111 neighbourhoods. Lombok is subdivided into 5 sub-neighbourhoods, while the historic inner city is subdivided in 8 sub-neighbourhoods. After testing for multicollinearity and checking the complete correlation matrix for all independent variables, we chose to include: 'the percentage of children between 0 and 15 years old'; 'the

Table 2. A multivariate linear regression analysis on the relative presence of Airbnb per neighbourhood.

Variables	Dependent variable: relative presence of Airbnbs (Airbnbs per 100 households) B (Beta)
Presence of families with children (% of children between 0 and 15 years old among neighbourhood population)	−0.058 (−0.340)**
Presence of non-Western minorities (% of neighbourhood inhabitants with a first or second generation non-Western background)	0.007 (0.079)
Presence of low-income households (% of households belonging to the 40% lowest household incomes in the Netherlands)	−0.024 (−0.315)**
Average housing value (average neighbourhood housing value estimated by tax agency in 1.000€)	7822E-5 (0.005)
Presence of services (average number of restaurants located within 1 km from neighbourhood inhabitants)	0,013 (0,559)***
Distance to city centre (average distance to the Dom church for neighbourhood inhabitants in metres)	0.000 (−0.268)**
*$P < 0.10$; **$P < 0.05$; ***$P < 0.01$	R^2 0.662 Intercept 3.466 *n* 86

Sources: Central Bureau for Statistics (2017); AirDNA (2017).

percentage of low-income households'; 'the percentage of ethnic minorities'; and 'average property value'. We also included important socio-economic and demographic neighbourhood characteristics as well as the 'average number of restaurants within 1 km' to show the vicinity of services important for tourists. We calculated the average distance to the Dom church for all neighbourhoods to factor in a centrality measure and included this in the multivariate regression analysis as a last variable. This analysis was performed through SPSS and the results shown in Table 2.

Data analysis

Utrecht's Airbnb sector reveals several characteristics summarized in Table 1 and Figure 1, respectively. In addition to general data and Airbnb-specific information relating to the entire city, Table 1 summarizes information concerning Utrecht's historic centre and Lombok. Overall, the city has 174,765 households, 9950 of which reside in the centre and a further 11,280 in Lombok. Both Lombok and the city as a whole have sizeable percentages of non-western migrants (around 22%).

Between August 2016 and October 2017, the number of active Airbnbs in the entire city expanded by nearly 80% from 1202 properties to 2156. The centre witnessed the largest increase (88%) while Lombok experienced a more modest growth (47%). Of the 1202 active properties in the city in August 2016, most were entire homes/apartments (860) while a further 329 were private rooms and only 13 shared rooms. The number of active Airbnbs in the city per 100 households was 0.69. This figure was lower than that in the city centre (2.22) but also Lombok (1.35). Significantly, by October 2017 the numbers of Airbnb units per 100 households had increased to 4.17 for the city centre, 1.96 for Lombok, and 1.23 for the city as a whole.

Figure 1 reveals an Airbnb cluster in Utrecht's historic centre, spilling over into some of the nearby pre-1930 residential neighbourhoods. The latter, are those experiencing gentrification, including Lombok (Permentier et al., 2008). Meanwhile, the traditionally wealthy eastern neighbourhoods register few Airbnbs. The post-1960 neighbourhoods (e.g. Kanaleneiland or Overvecht), located beyond the centre – albeit connected to this via public

Figure 1. Relative distribution of Airbnb accommodations in Utrecht per neighbourhood active between October 2016 and October 2017. Source: AirDNA (2017); Central Bureau for Statistics (2017).

transportation – with high percentages of social housing and a noticeable population of ethnic minorities, have few such properties. This is also the case in family-oriented suburban neighbourhoods like Lunetten and within the city's recently expanded area where there is a high proportion of private homes and a low population of ethnic minorities (e.g. Leidse Rijn).

In the period ending on 22 October 2017, 140,000 overnights were recorded in the city's 2156 active Airbnbs. Indeed, 110,000 overnights were in the 832 properties described as 'very active'. These figures show an impressive increase from the period ending in August 2016 when 41,154 overnights were recorded in 1202 active properties and the total number of very active Airbnbs for the entire city was only 199. Very active Airbnbs are spread throughout the city. Both within the city centre and in Lombok they constitute around 39% of the entire stock. Between August 2016 and October 2017, there was an increase in the so-called 'professional' Airbnbs for the city as a whole from 7.8% to 9.9% of the total. This means that about one tenth of the capacity represents properties owned by hosts with multiple listings who make these available for more than 2 months in a calendar year. Both in the centre and in Lombok the respective proportion of professional listings for the period ending 22 October 2017 was higher than the city-wide average.

Considering that only a few years ago there were almost no places to stay overnight in Lombok the appearance of 224 accommodation units constitutes a considerable transformation (Figure 2). Just like the city centre, Lombok has seen an increase in the booking

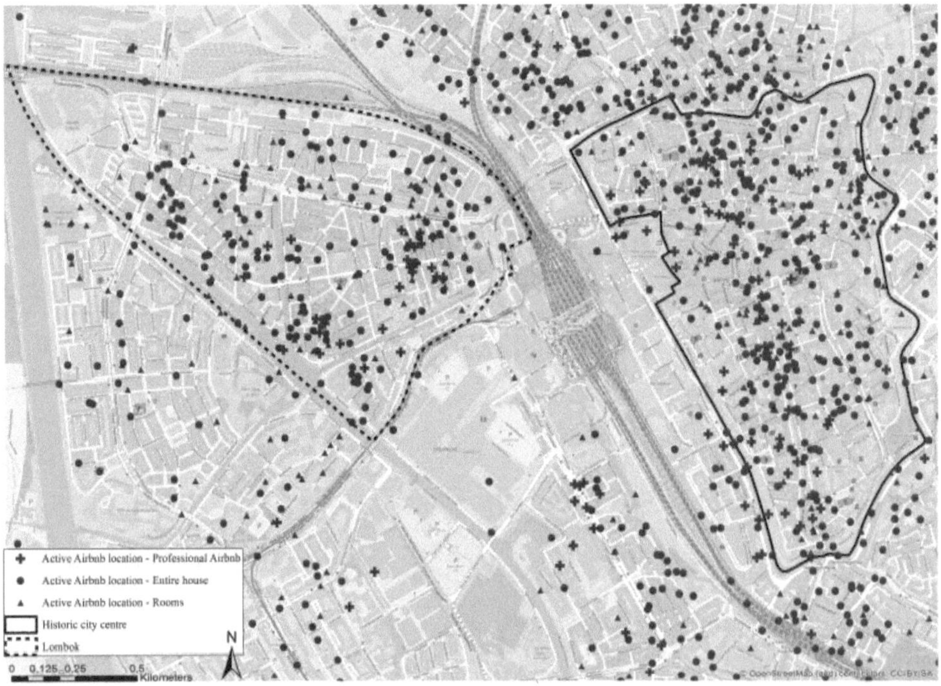

Figure 2. Distribution of Airbnb accommodation in central Utrecht active between October 2016 and October 2017. Source: AirDNA (2017); OpenStreetMap contributors (2017).

numbers and nights per active Airbnb. The average number of nights per active Airbnb has doubled in the period August 2016 to October 2017, now standing at 69.4.

To ascertain the factors influencing the location of Airbnb in Utrecht and to see the relationship between this activity and other variables we performed a cross-neighbourhood study. We examined 86 out of the city's 111 neighbourhoods by conducting a multivariate regression to detect neighbourhood characteristics influencing Airbnb presence. Neighbourhoods with missing data (probably because they are non-residential or have small populations) were omitted, while we excluded one of the central city neighbourhoods as a statistical outlier; although this area has very few residents, AirDNA indicates the presence of a high number of Airbnbs due to location distortion.

There is a distance decay effect of Airbnb activity as one moves outward from the city centre (defined as the Dom church) (Table 2). Unsurprisingly, the highest concentration is in neighbourhoods within and surrounding the city centre where the central station and most of the visitor attractions and tourist infrastructure are located. The statistical analysis demonstrates that the number of Airbnbs is highest in neighbourhoods with restaurants within 1 km. Indeed, the presence of restaurants within close proximity is the strongest explanatory variable for Airbnb presence. In Lombok itself, the fact that Airbnb is clustered mostly in the neighbourhood's western part closest to the centre reinforces the influence that services such as eateries have on such activity. There is also a negative correlation between lower-income neighbourhoods and Airbnb presence. Interestingly, neighbourhoods with high proportions of children under 15 years of age have few such properties. To summarize, the multivariate analysis indicates that Airbnbs are likelier to be present in neighbourhoods offering a number of touristic services (such as restaurants), which are near the city centre and where there are few families with children and also few low-income households. The presence of ethnic minorities and also property values do have an effect on Airbnb presence. Overall, these results are not in the least surprising, considering that in areas that already draw many visitors by virtue of their attractions and services on offer enterprising owners of Airbnb properties can command high prices but also occupancy rates. Their opportunity for profit-making is far greater than it would be in a leafy suburban residential neighbourhood.

Discussion and conclusions

Although Utrecht is a historic mid-sized city with several attractions and despite drawing a considerable number of visitors (mostly on day trips) it has not yet emerged as an over-touristified hotspot compared to similar European cities (e.g. Bruges and Oxford). This is partly true because Utrecht's economy in quite diversified (it has a major university and serves as a major railway hub in the Netherlands). This situation made it ideal for investigating how Airbnb influences communities that are not yet major tourism magnets but are rapidly emerging as the 'must see' hidden gems of global tourism.

Overall, Airbnb activity remains at its early stages with few active properties per 100 households compared to far more popular venues (e.g. Venice or Reykjavik). Nevertheless, despite these modest figures, the number of Airbnb units is increasing rapidly and in 2016, this option represented roughly 15% of total overnight stays in the city. Indeed, given the city's shortage of hotel accommodation, an alderman of Utrecht was quoted in a national newspaper article (Franck, 2016) saying 'Airbnb complements the offer of touristic accommodation in Utrecht'.

Our study reveals that just as in more heavily touristified venues, Utrecht's Airbnb activity is growing, especially within the city centre and immediate surroundings. This suggests that the transformation of regular housing into Airbnb rentals strongly relates to nearby tourism attractions but also supporting services including restaurants and hotels. This observation, in turn, reinforces findings from Barcelona demonstrating that rather than spreading evenly throughout the city, Airbnbs cluster in neighbourhoods in and around the city's core, within close proximity to hotels (Arias Sans & Quaglieri Domìnguez, 2016).

There is an obvious distance-decay effect from the very centre of the city (marked by the Dom Tower) in the number of active Airbnb rentals. As the city itself witnesses increasing Airbnb activity, gradually this activity spreads to surrounding areas all of which are within easy access to the central station, the main visitor attractions, and facilities including restaurants.

Focusing specifically on Lombok, which displays neo-bohemian traits – albeit at an early stage – Airbnb is not evenly spread throughout the neighbourhood but rather occurs in the part closest to the city centre, which also happens to be the area where most of the touristification is occurring. Most Airbnbs are concentrated within the neighbourhood's western edge, an area where the higher-end restaurants and upscale cafes have been appearing. Further into its core, Lombok retains its ethnic character and there is a noticeable decrease in Airbnb activity. This suggests that Airbnb activity is a process following after the initial intensification of tourism activities in particular areas and does not occur haphazardly throughout a city. In other words, individuals owning one or more property are likelier to list these through the Airbnb platform when located in or close to established visitor areas.

We could argue that geography matters since in the case of Lombok, an emerging neo-bohemia, the chances for local resources (i.e. homes) to become monetized and attractive as tourism business investments (i.e. Airbnb or other short-term rentals) depend highly on physical proximity to mainstream tourist attractions but also, importantly, formal tourism infrastructure (i.e. hotels, restaurants, bars, shops). This is especially the case for 'professionally driven' properties (ones controlled by multi-property owners who lease them out over two months per year). Thus, proximity to the existing city centre tourist bubble appears to be a strong explanatory variable in determining Airbnb density in Utrecht and as activities from this bubble spill over into surrounding areas, including Lombok, Airbnb follows. Also, as opposed to Guttentag (2015, 2016, 2017) who describes Airbnb as an opportunity for visitors to seek out and experience a variety of local and authentic spaces (MacCannell, 1999), the appeal as to where to stay seems to be more hedonic and utilitarian in nature. This likely explains why in Lombok the demand for Airbnb stays and the average income generated from such units is higher than the city-wide average with the exception of listings located within the existing city-centre bubble.

Whether Airbnb causes any tangible or intangible negative effects on, for example, the housing market, the hotel industry, or the neighbourhood's social fabric is far from certain at this moment. Digging into this issue in future studies, would be most relevant given that, for instance, recent gentrification processes in Utrecht have already led to expanding conflict between the well-educated and high-income newcomers into Lombok and the long-term residents and shopkeepers (Huisman, 2017). Indeed, the area is already quite fragile and the proliferation of Airbnb activity as the trends suggest could signify yet another layer of gentrification processes that might eventually prove harmful to Lombok's social fabric. As demonstrated by Lee (2016), the likelihood is high that owners transform property into short-term rental units as soon as they have proof that this is a more profitable avenue than making their homes available to the long-term housing market. In that sense, geographical location, that can either be thought of as an advantage or a disadvantage, may in the end invite for more tensions and, subsequently, lead to further scholarly criticism being aimed at Airbnb (Fang et al., 2016; Gant, 2015; Interian, 2016; Lee, 2016; Lehr, 2015; Neeser, 2015; Zervas & Byers, 2016).

Undoubtedly, Lombok is experiencing a gradual transformation into a neo-bohemian district increasingly catering to new urban tourism. This transformation, however, is not happening uniformly within the neighbourhood nor is it always overtly visible. To be sure, there are transformations of second-hand shops into designer boutiques catering to an upscale clientele, refurbished factory buildings metamorphosing into restaurants aimed at a cosmopolitan class (Ioannides & Petridou, 2016), or the introduction of global brands such as the hotel-chain Ibis, located close by in downtown Utrecht. Nevertheless, transforming residential units into Airbnb within Lombok is both a clandestine and camouflaged process whereby a multinational presence has began infiltrating neo-bohemia. Thus, although we could consider Lombok as a space without obvious marks of the city centre bubble - it is less regulated and predictable and bears few, if any, signs of multinational capital (Edensor, 1998; Fainstein et al., 2003; Fuller & Michel, 2014) - this neighbourhood shows signs of its emergence as a place of cosmopolitan and multinational production and consumption. A major force behind this shift is Airbnb as a global brand.

This occurs in a neo-bohemian context that outwardly projects the illusion of 'edginess' where non-scripted experiences associated with predictable bubbles are shunned (Ioannides & Petridou, 2016; Judd, 2003; Lloyd, 2006; Maitland, 2007). It happens in an area where just recently, a former simple kebab-store owned by an immigrant was upgraded into a classy Neapolitan pizzeria despite the fact that the premises kept their 'edgy' décor so as not to stand out too much from the surroundings (DUIC, 2017).

Ultimately, these processes suggest that 'mundane' neighbourhoods like Lombok, which are transforming into neo-bohemian spaces also project signs of Airbnb intensification. This phenomenon arises as individual homeowners but also real-estate barons increasingly recognize the profit-making capabilities of their properties. Further Airbnb activity and that of other short-term rental platforms in the neighbourhood is likely to lead to even further visits and cause intensification of tourist-related activities in the form of additional restaurants, nightclubs, upscale cafes and related venues. The gradual touristification of parts of Lombok, which is likely to eventually be accompanied by an exodus of some long-term residents as more properties transform into Airbnbs could signify the expansion of the nearby centre-city tourism bubble. To be sure, we cannot assume from this one case study that this is a phenomenon that is repeated in every city but there is enough evidence to suggest that in places like Utrecht where there is a nascent touristification process, in the absence of a clearly targeted policy related to short-term rentals, certain neighbourhoods run the risk of being transformed into bubbles in their own right. A final point to ponder over is whether such a developmental path will enable a neighbourhood like Lombok to achieve a healthy balance whereby experience-seeking visitors coexist with multiple groups of locals, or whether never-stopping gentrification processes will eventually kill the community's spirit.

Disclosure statement

No potential conflict of interest was reported by the authors.

ORCID

Dimitri Ioannides http://orcid.org/0000-0002-3549-750X

References

AirDNA. (2016). Property report Utrecht August 2016 [data-set]. Retrieved from https://www.airdna.co/

AirDNA. (2017). Property report Utrecht October 2017 [data-set]. Retrieved from https://www.airdna.co/

Arias Sans, A., & Quaglieri Dominguez, A. (2016). Unravelling Airbnb: Urban perspectives from Barcelona. In A. P. Russo & G. Richards (Eds.), *Reinventing the local in tourism: Producing, consuming and negotiating place* (pp. 209–228). Bristol: Channel View Publications.

Baranjuk, C. (2017). The country that tourism has taken by surprise. Retrieved from http://www.bbc.com/future/story/20170222-the-country-that-tourism-has-taken-by-surprise

Belk, R. (2010). Sharing. *Journal of Consumer Research, 36*(5), 715–734.

Central Bureau for Statistics. (2017). CBS in Uw Buurt [data-set]. Retrieved from http://www.cbsinuwbuurt.nl/#buurten2016_aantal

Colomb, C., & Novy, J. (Eds.). (2017). *Protest and resistance in the tourist city*. London: Routledge.

Comiteau, L. (2016). Is Airbnb making it impossible to find a house? Retrieved from http://www.bbc.com/capital/story/20160811-is-airbnb-making-it-impossible-to-find-a-house

Dredge, D., & Gyimóthy, S. (2015). The collaborative economy and tourism: Critical perspectives, questionable claims and silenced voices. *Tourism Recreation Research, 40*(3), 286–302.

Dredge, D., & Gyimóthy, S. (2017). *Collaborative economy and tourism: Perspectives, politics, policies and prospects*. Cham: Springer.

Dudás, G., Vida, G., Kovalcsik, T., & Boros, L. (2017). A socio-economic analysis of Airbnb in New York City. *Regional Studies, 7*(1), 135–151.

DUIC. (2017). Nieuwe pizzeria in Lombok: 'Punk Pizza moet leuk zijn voor de buurt'. Retrieved from https://www.duic.nl/horeca/nieuwe-pizzeria-lombok-punk-pizza-moet-leuk-zijn-voor-de-buurt/

Edensor, T. (1998). *Tourists at the Taj. Performance and meaning at a symbolic site*. London: Routledge.

European Commission. (2016). *A European agenda for the collaborative economy*. Brussels: European Commission. Retrieved from http://eur-lex.europa.eu/legal-content/EN/TXT/?uri=COM%3A2016%3A356%3AFIN

Fainstein, S. S., Hoffman, L. M., & Judd, D. R. (2003). Making theoretical sense of tourism. In L. M. Hoffman, S. S. Fainstein, & D. R. Judd (Eds.), *Cities and visitors: Regulating people, markets, and city space* (pp. 239–253). Oxford: Blackwell Publishing.

Fang, B., Ye, Q., & Law, R. (2016). Effect of sharing economy on tourism industry employment. *Annals of Tourism Research, 57*, 234–278.

Forno, F., & Garibaldi, R. (2015). Sharing economy in travel and tourism: The case of home-swapping in Italy. *Journal of Quality Assurance in Hospitality & Tourism, 16*(2), 202–220.

Franck, R. (2016). Airbnb'ers positief over plannen van de gemeente. AD Nieuws. Retrieved from http://www.ad.nl/utrecht/airbnb-ers-positief-over-plannen-van-de-gemeente~acf15812/

Füller, H., & Michel, B. (2014). 'Stop being a tourist!' New dynamics of urban tourism in Berlin-Kreuzberg. *International Journal of Urban and Regional Research, 38*(4), 1304–1318.

Gant, A. (2015). Tourism and commercial gentrification. Retrieved from http://www.rc21.org/en/wp-content/uploads/2014/12/E4-C%C3%B3cola-Gant.pdf

Gotham, K. F. (2005). Tourism gentrification: The case of new Orleans' vieux carre (French Quarter). *Urban Studies, 42*(7), 1099–1121.

Guttentag, D. (2015). Airbnb: Disruptive innovation and the rise of an informal tourism accommodation sector. *Current Issues in Tourism, 18*(12), 1192–1217.

Guttentag, D. (2016). *Why tourists choose Airbnb: A motivation-based segmentation study underpinned by innovation concepts* (Doctoral dissertation). University of Waterloo, Waterloo

Guttentag, D. (2017). Regulating innovation in the collaborative economy: An examination of Airbnb's early legal issues. In D. Dredge & S. Gyimóthy (Eds.), *Collaborative economy and tourism: Perspectives, politics, policies and prospects* (pp. 97–128). Cham: Springer.

Gutierrez, J., Garcia-Palomares, J. C., Romanillos, G., & Salas-Olmedo, M. H. (2017). Airbnb in tourist cities: Comparing spatial patterns of hotels and peer-to-peer accommodation. *Tourism Management, 62*, 278–291.

Gyimóthy, S. (2017). Definitions and mapping the landscape in the collaborative economy. In D. Dredge & S. Gyimóthy (Eds.), *Collaborative economy and tourism: Perspectives, politics, policies and prospects* (pp. 15–30). Cham: Springer.

Hackworth, J. (2007). *The Neoliberal city: Governance, ideology, and development in American urbanism*. Ithaca, NY: Cornell University Press.

Hoffman, L. M., Fainstein, S. S., & Judd, D. R. (Eds.) (2003). *Cities and visitors: Regulating people, markets, and city space*. Oxford: Blackwell Publishing.

Horwath. (2014). *Onderzoek Hotelmarkt Utrecht*. Amsterdam: Horwath HTL.

Horwath. (2016). *HOSTA 2016*. Amsterdam: Horwath HTL.

Huisman, C. (2017). Wat doet de yup met Lombok? *De Volkskrant*, 12–13.

Inside Airbnb. (2017). Retrieved from http://insideairbnb.com/about.html

Interian, J. (2016). Up in the air: Harmonizing the sharing economy through Airbnb regulations. Retrieved from http://lawdigitalcommons.bc.edu/iclr/vol39/iss1/6/

Ioannides, D., & Petridou, E. (2016). Contingent neoliberalism and urban tourism in the United States. In J. Mosedale (Ed.), *Neoliberalism and the political economy of tourism* (pp. 21–35). New York, NY: Routledge.

Johnson, C. (2015). Gentrifying New Orleans: Thoughts on race and the movement of capital. *Souls, 17*(3–4), 175–200.

Judd, D. (2003). Visitors and the spatial ecology of the city. In L. M. Hoffman, S. S. Fainstein, & D. R. Judd (Eds.), *Cities and visitors: Regulating people, markets, and city space* (pp. 23–38). Oxford: Blackwell Publishing.

Kathan, W., Matzler, K., & Veider, V. (2016). The sharing economy: Your business model's friend or foe? *Business Horizon, 59*(6), 663–672.

Lee, D. (2016). How Airbnb short-term rentals exacerbate Los Angeles's affordable housing crisis: Analysis and policy recommendations. *Harvard Law & Policy Review, 10*, 229–253.

Lehr, D. (2015). *An analysis of the changing competitive landscape in the hotel industry regarding Airbnb* (Master's thesis). Dominican University of California, San Rafael, CA.

Lloyd, R. (2006). *Neo-Bohemia: Art and commerce in the postindustrial city*. New York, NY: Routledge.

Lonely Planet. (n.d.). Ten of the world's unsung places. Retrieved from https://www.lonelyplanet.com/travel-tips-and-articles/10-of-the-worlds-unsung-places/40625c8c-8a11-5710-a052-1479d277e7b8

MacCannell, D. (1999). *The tourist: A new theory of the leisure class*. Berkeley, CA: University of California.

Maitland, R. (2007). Tourists, the creative class and distinctive areas in major cities: The roles of visitors and residents in developing new tourism areas. In G. Richards & J. Wilson (Eds.), *Tourism, creativity and development* (pp. 73–86). New York, NY: Routledge.

Molz, J. G. (2013). Social networking technologies and the moral economy of alternative tourism: The case of couchsurfing.org. *Annals of Tourism Research, 43*, 210–230.

Neeser, D. (2015). *Does Airbnb hurt hotel business: Evidence from the Nordic countries* (Master's thesis). Universidad Carlos III de Madrid, Madrid.

OpenStreetMap contributors. (2017). Utrecht [Map]. Retrieved from https://www.openstreetmap.org/search?query=utrecht#map=12/52.0810/5.0997

Oppilard, F. (2017). From San Francisco's 'Tech Boom 2.0' to Valparaiso's UNESCO World Heritage Site: Resistance to tourism gentrification from a comparative political perspective. In C. Colomb & J. Novy (Eds.), *Protest and resistance in the tourist city* (pp. 129–151). London: Routledge.

Oskam, J., & Boswijk, A. (2016). Airbnb: The future of networked hospitality businesses. *Journal of Tourism Futures, 2*(1), 22–42.

Permentier, M., Van Ham, M., & Bolt, G. (2008). Same neighbourhood… different views? A confrontation of internal and external neighbourhood reputations. *Housing Studies, 23*(6), 833–855.

Sigala, M. (2015). Collaborative commerce in tourism: Implications for research and industry. *Current Issues in Tourism, 20*(4), 1–20.

Sundararajan, A. (2015). The 'gig economy' is coming. What will it mean for work?. Retrieved from https://www.theguardian.com/commentisfree/2015/jul/26/will-we-get-by-gig-economy/

Toerisme Utrecht. (2015). *Meerjarenstrategie vrijetijdseconomie 2015–2020*. Utrecht: Toerisme Utrecht.

Tussyadiah, I. P., & Pesonen, J. O. (2015). Impacts of peer-to-peer accommodation use on travel patterns. *Journal of Travel Research, 55*(8), 1022–1040.

Wood, B. (2017). Airbnb is now bigger than the world's top five hotel brands put together. Retrieved from https://thespaces.com/2017/08/15/airbnb-now-bigger-worlds-top-five-hotel-brands-put-together/

Yrigoy, I. (2016). The impact of Airbnb in the urban arena: Towards a tourism-led gentrification? The case study of Palma old quarter. In M. Blazquez, M. Mir-Gual, M. Murray, & G. X. Pons (Eds.), *Turismo y Crisis. Turismo Colaborativo y Ecoturismo*. XV Coloquio de Geografia del Turismo, el Ocio y la Recreacion de la AGE. Mon. Soc. Hist. Nat. Balears, 23, pp. 281–289.

Zervas, G., & Byers, J. (2016). The rise of the sharing economy: Estimating the impact of Airbnb on the hotel industry. Retrieved from http://cs-people.bu.edu/dproserp/papers/airbnb.pdf

A place in the sun: the British housing market's (de)construction of Andalusia

Miguel García Martín, Arsenio Villar Lama and Estrella Cruz Mazo

ABSTRACT

Andalusia has become one of the main destinations for British residential tourists, attracted by leisure, a mild climate and the search for a better lifestyle. Prompted by its interest in selling houses, the real estate industry has devised an image of the Andalusian landscape that is, to all intents and purposes, partial and biased. The purpose of this research is to assess the importance of the elements and features that characterise this particular vision of this Spanish region, paying particular attention to more rural inland areas. For this, an empirical study has been conducted based on content analysis of real estate websites targeting British buyers. This study has revealed that: (i) the misrepresented and romantic projection of the image of Andalusia still persists to a certain extent. Among the main elements and attributes that characterise the landscape are all those that emphasise the region's sublime, picturesque and exotic nature; other main landscape qualities and unique areas, specifically, any that today enjoy natural or cultural legal protection due to their heritage value, are less important in this cultural construction; (ii) the various parts of Andalusia are treated unequally. Malaga province and the rest of the Baetic Mountains are much better represented than other areas (the Guadalquivir Valley and the Sierra Morena mountains). This imbalance sometimes results in a distorted construction of the territory. This way of conceiving Andalusia puts Malaga at the heart of the region, and the further from the Costa del Sol, the more peripheral the regions become; (iii) the complexity of the landscape as a cultural expression is reduced to a set of simplified features that can be easily digested by a social group focused not on experiencing, but on consuming the landscape.

1. Introduction

On a popular British television show, a couple of home buyers try to find their ideal property in a sun-blessed foreign country. While the couple weigh up the pros and cons of purchasing either a whitewashed country house in Axarquía (southern Spain) or a detached villa with pool on the Algarve (Portugal), the TV programme presenter helps them to make the best choice. This programme, called A Place in the Sun, combines a real estate reality show and a television travel series. It offers a very particular image of the Southern Europe landscape seen through the shrewd eye of the housing market and the spectacle culture, as interpreted by Kellner (2004). One might think that, for all those comfortably sat on the couch in their homes, neither Andalusia nor the Algarve is anything more than a sunny place to own a holiday home or retire to.

In fact, in recent decades, the British and other Northern Europeans have developed a residential model based on owning a home in Southern Europe. This flow of permanent, semi-permanent (seasonal) or occasional visitors has driven the growth of urban and suburban surfaces in coastal and inland areas to meet the demand for second or new homes. Andalusia, the southernmost region of mainland Spain, has become a major destination for British visitors and migrants. While coastal areas have been the phenomenon's traditional target area, as a sun and sand climate destination for their beaches, inland Andalusia and its rural countryside are starting to gain in importance. The real estate industry has reacted by developing a carefully constructed projected image of the territory to drive house sales and command a presence in a competitive market.

As a consequence, the industry is transmitting a particular vision and idea of Andalusia, a contrived image that is, to all intents and purposes, partial and biased, and driven by profit-oriented interest: in the commodification of the rural landscape, some particular and positive features are intentionally highlighted, whereas others are totally and utterly ignored. Thus, the purpose of this paper is to assess the importance of the features that characterise this peculiar vision of Andalusia, paying particular attention to the more rural inland territories. For this, an empirical study has been conducted, based on content analysis of real state websites aimed at British buyers.

Landscape should be considered a social and cultural construct (Cosgrove, 1984; Maderuelo, 2005), rather than the mere identification of an area of territory or even its perception. Throughout the 19th and the beginning of the 20th centuries, British and other romantic travellers played a significant role in the construction of this specific, contrived image of Andalusia, which is depicted as mythic and exotic in the travel literature: sublime beauty, untamed nature, ruins, Moorish heritage and passionate people construct a picturesque portrait that has been perpetuated throughout subsequent decades (López-Ontiveros, 1988). This literary genre was a fundamental precedent in the shaping of Spanish mass tourism culture throughout the twentieth century. Richard Ford's written legacy is a good example: as a pioneer British travel writer, he recorded his experience as an observer and discoverer, describing the landscapes, customs and characters of the various Spanish regions. His influential *A Handbook for Travellers in Spain*, first edited in 1845, was more than a simple guide: it was a compendium of the art of travel intended for an ideal traveller (Pillet, 2016; Robertson, 2014). The success of this type of book contributed conclusively to the construction of a mythological and picturesque image of Spain among foreign audiences.

The inclusion of the region in the mass tourism phenomenon during the last third of the 20th century reshaped the projected and perceived image of Andalusia that has remained to the present day: This way of experiencing a place spread from an exclusive and cultural elite to mainstream consumers, i.e. blue collar workers from Northern Europe (Storm, 2013). The sense of the famous slogan, "Spain is different", is essential for understanding how deeply the country has taken hold as a leading tourist destination, particularly for British visitors. Spain became a major tourism power over the course of two decades, between the 1960s and 1970s. The Franco regime saw in the tourist industry a useful vehicle to stabilize the domestic economy. But it was also harnessed as a tool for political self-legitimization in the eyes of European democracies and a way to strengthen the country's international presence in all these countries (Pack, 2009).

Thus, a new relationship between the British (and other North European nationalities) and Spanish destinations was forged, mediated by a cheap 'sun and sand' offer. The rapid emergence of a number of coastal resorts specifically targeted at the British in enclaves along the Mediterranean fringe led to a something akin to 'mini colonies'. To a certain extent, tourists are detached from the local physical and cultural environment in these places, which only serves to reinforce their sense of belonging to their own national communities (Andrews, 2011).

This significant bond between the UK and Spain in tourism terms has been strengthened in recent years (cf. a geographical approach in Section 3); Spain is the top destination for Britons, and they represent the largest group of foreign visitors to Spain.

This paper is structured as follows: firstly, the main concepts and ideas are addressed through a literature review. A contextual approach is also presented: the topic is framed by an analysis of general aspects and data on demographics, tourism and housing investment. Secondly, the methodology is described. This consists of an empirical implementation of qualitative techniques through housing promotion on the Internet. Next, the main results are presented. Finally, the principal conclusions are set out and possible lines of future research outlined.

2. Literature review

Mediterranean coastal areas have experienced a demographic inflow from other European countries in recent decades, mostly, but not entirely, of retirees, who mainly come from Northern and Western Europe and are attracted by leisure, good weather and the search for a better lifestyle.

Academic approaches and studies in the field have provided several definitions of this complex phenomenon, highlighting the close links between tourists, migrants, retirees, residents and home ownership. Despite difficulties in marking the conceptual boundary between the two geographical realities, some authors opt for the residential tourism concept, stating that tourism destinations are a primary stage in the consolidation of residential mobility (Rodríguez, 2001; Williams & Hall, 2000). Retiree status is a third vector in the composition of this kind of global dynamic, to the extent that international retirement migration is usually represented by its abbreviation, IRM (Warnes, 2009; Williams, King, & Warnes, 1997). Benson and O'Reilly (2009) refer to the notion of lifestyle migration, arguing that other conceptualisations fail to be fully inclusive. Some researchers draw the fuzzy boundaries of residential tourism by classifying it into categories such as: production- or consumption-related; registered or unregistered; lifestyle-oriented; temporary or permanent mobility, etc. (Bell & Ward, 2000; Cohen, Duncan, & Thulemark, 2015). This latter contribution establishes the different patterns of these three types of contemporary mobility as a crossroads between travel, leisure and migration. Recently, Mantecón (2017) has discussed whether the concept of residential tourism is an oxymoron, a concept made up of two terms that allude to contradictory realities (tourism and residence), in contrast to its perfectly coherent real estate label.

Motivations have been fully analysed in the literature by surveying new residents (see Casado-Díaz, Kaiser, & Warnes, 2004; Huete & Mantecón, 2012; O'Reilly, 2007): climate; improvement to health; financial conditions (cost-of-living, housing advantages); lifestyle and culture (local life, cuisine, wine, outdoor leisure); proximity to their country of origin; environmental conditions; prior connection with the place; and proximity to fellow citizens, among other things. In addition, these authors usually identify some push factors—getting away from their country of origin—like poor weather, stress, high tax and/or crime rates, and crucial life events, being this latter the so-called 'triggers' (Breuer, 2005). The relative importance of these factors differs by nationality, socio-economic group and location. In the case of the Costa del Sol (as the main Andalusian area under study), the search for good weather—often linked to treating an illness—and the 'Spanish way of life' seem to be the two main motivations. In relation to the latter, although lifestyle is considered a key motive, many British residents appear to be under integrated or even excluded from local life (Casado-Díaz, 2006; Huete & Mantecón, 2013; O'Reilly, 2007).

Whichever of the terms is accepted, and leaving personal motivations aside, one of the key factors that puts these migrations into context is the existence of an internationalised real estate market that converts the visitor into a homeowner. Thus, tourism and migration can be understood as intertwined scopes and the second home the point of contact (Müller, 2002a, 2002b). Williams, King, and Warnes (2004) stress this close link, observing how the second home—the house itself, its form of ownership, the buying and selling process, flows of property between the country of origin and the country of destination, etc.—determine the life stories of the tourists/residents. In fact, these last authors propose some more accurate designations—'alternative' or 'multiple' house instead of second home—in view of the fact that 'an increasing share of the population is adopting residential strategies that involve complex sequential shifts in the functional use of multiple properties' (Williams et al., 2004, p. 112).

Aside from theoretical approaches and terminological contributions, a major part of academic work focuses on empirical methodologies based on study cases in specific regions of Southern Europe. Far from observing the phenomenon from a structural non-specific perspective, several studies have concentrated on very concrete cases concerning specific nationalities and destinations: the British on the Costa del Sol (O'Reilly, 2000, 2017), Northern Europeans on the Costa del Sol (Olson & O'Reilly, 2017; Rodríguez, Fernández-Mayoralas, Rojo, & Abellán, 2005) and the Costa Blanca (Casado-Díaz, 2006; Huete & Mantecón, 2013; Mazón, 2006); or on comparing destinations with similar environmental and socio-cultural patterns: the British in Tuscany, Malta, the Algarve and the Costa del Sol (King, Warnes, & Williams, 2000; Warnes, King, Williams, & Patterson, 1999). Apart from analysing these groups economically and demographically, one fundamental motivation for migration that these contributions highlight is the migrants' search for the better quality of life that a more benign climate provides. They also underline their satisfaction with their new way of life, despite significant, but not particularly bothersome, integration constraints (due to the language barrier). In addition, Huete, Mantecón, and Mazón (2008) shed light on perceptions of the host society through a qualitative analysis of local stakeholder discourses.

Other authors have focused on the territorial consequences of residential tourism, with urbanisation and urban sprawl standing out on the Costa Blanca (Mazón & Aledo, 2005; Membrado, 2015). An absence of adequate urban or tourism planning and the link with speculation processes seem to be the most characteristic keys of this scope. Mantecón (2017, p. 696) provides a fresh definition of residential tourism that directly ties in with this scope: 'a label integrated in a publicity strategy at the service of political and business interests linked directly to the real estate economy'.

Meanwhile, a research line has emerged that is focused on addressing the phenomenon of return migration to Britain (see Betty & Hall, 2016; Giner-Monfort, Hall, & Betty, 2016). This is a subject that has received less academic attention, despite the notable changing context of the deceleration, or even decrease, of the migratory balance in Spain. Lack of health or social care, the loss of personal and institutional support and a new economic outlook (the depreciation of the pound sterling, the increased cost of living in Spain, etc.) might help explain the intention of an increasing number of British residents to return to their country of origin. Hall and Hardill (2016) underline the vulnerability of certain elderly citizens and the responsibilities of the British and Spanish Administrations, with this group of migrants under-represented in policy making.

Other approaches have addressed the Northern European population in Southern Mediterranean areas from a widespread socio-cultural focus: from the involvement of migrants in local political life (Janoschka & Durán, 2014) to new environments generated by the ornamental plants (especially palm trees) found in recent urban developments in Alicante (Hernández, Morales, & Sauri, 2014). The historical perspective of British expatriates in Pre-Civil War Spain has even been addressed (Shelmerdine, 2002).

However, the majority of these scientific contributions can be sorted into two main groups: the first dealing with migrants' social experiences: their motivations, decision-

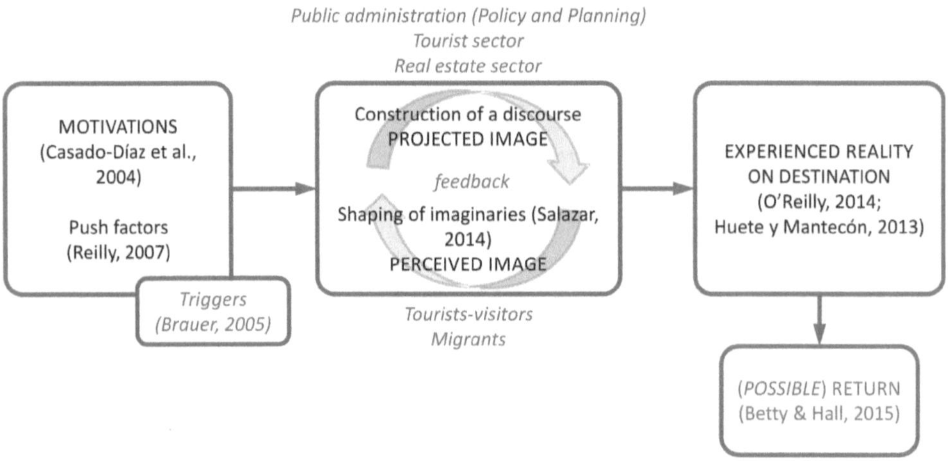

Figure 1. Lifestyle migration sequence: Motivations–social imaginaries–experienced reality. Referred authors have been considered as representative of every topic. Source: Authors.

making, expectations, preferences and likes, etc. (the sociological aspect, focused on people); and the second, with the spatial consequences of the migration process: urban growth, land use and landscape changes, the evolution and expansion of second homes, ecological impacts, etc. (the environmental aspect, focused on the territory).

The two research lines merge in the lifestyle migration sequence (Figure 1), where motivations, construction-perception of an image, experienced reality and returns are part of the chain. The literature that addresses the socio-cultural representation of nature and the construction of an image of rurality (in which this contribution is set) sits half-way between social experiences and environmental approaches. This has been well-documented in the case of German and Dutch second home owners in rural Sweden (Eimermann, Lundmark, & Müller, 2012; Müller, 2002a, 2002b) and the British in rural France (Benson, 2011; Hoggart & Buller, 1994; Lord, 2008). According to Benson, the landscape plays a core role in a better way of life. Beautiful scenery and the promise of a traditional rural life style (the so-called rural idyll) provide a sense of uniqueness that cannot be found in the coastal mass tourism destinations. Indeed, in these kinds of inland destinations, migrants demonstrate greater commitment to, and integration in, both the physical and the social host environment.

Nevertheless, migration based on a search for a rural idyll has not been the target of any significant scientific attention in the Spanish and (especially) Andalusian cases. Apart from the work of Bertuglia, Sayadi, López, and Guarino (2013) on international *counterurbanisation* in the Alpujarra region, and the contribution of Cáceres-Feria and Ruiz (2017) on neo-rurality in the Sierra de Huelva, this academic scope continues to be a neglected research topic.

At this point, one might ask why the academic literature has overlooked the landscape as an important factor in tourist or migrant decision-making. Some considerations can be drawn from the previous research review: on the one hand, most sociological scopes could be explained by the predominant use of survey-based consolidated quantitative methodologies. Thus, to give an example, it seems easier to ask new residents about their motivations for moving, becoming involved in local or political life or preferences as to house types than on aspects that are difficult to verbalize, such as landscape perceptions and judgements. Similarly, territorial approaches rely on spatial analysis methods to process large amounts of information (land occupation, demographic data, etc.).

However, the landscape and the rural idyll are harder to treat as social constructions—shaped by different social practices and diverse economical and civil agents—in applied research beyond the plentiful literature and long-lasting epistemological debates. Benson (2011) is a case in point *t*: this is an in-depth investigation, involving close contact with new residents through face-to-face interviews and other qualitative techniques. Although this type of research is not very feasible in methodological terms as it requires a great deal of time, effort and resources, the results are extremely enriching. The present paper fills this apparent gap in this regard, as it uses a method that overcomes these material difficulties.

3. Mapping British house buyers: a geographical approach

There are 1.2 million British people living in other European Union countries; Spain hosts the largest community ahead of Ireland and France, with nearly a quarter of these (308,821). On the worldwide scale, Spain is the fourth-ranked hosting-country for UK citizens, behind Australia, the United States and Canada (United Nations, 2015). According to Instituto Nacional de Estadística—the Spanish National Statistics Office—(INE, 2018), the British are the third largest community in Spain after Romanians and Moroccans (both of which stand at just under 700,000 inhabitants). It is the largest group as far as so-called lifestyle migration is concerned, above other countries such as Italy (approx. 190,000), Germany (142,000) and France (101,000).

These numbers might be framed in the immigration movement from Central and Northern Europe, especially the retirees who have purchased residential properties in certain coastal locations around the Mediterranean Basin due to a number of different factors: climate comfort, house pricing, Spain's tourism status, citizen and legal security, health and food safety, EU integration and transport development in both flight connectivity and road accessibility (Fernández & Cruz, 2011).

As Figure 2 shows, the UK resident phenomenon is focused on the Mediterranean area, especially the southeast fringe between the Costa del Sol and the Balearic Islands, and is particularly intense in the provinces of Alicante and Malaga, with 74,400 and 51,900 British inhabitants respectively, equivalent to 4% and 3% of the total regional populations (INE, 2018). A first glance the map illustrates the relative importance of coastal areas compared to the rest of the country. However, a closer look allows a finer analysis: apart from the Costa del Sol and its surroundings (the pre-coastal zone), British communities are spread over other Eastern-Baetic areas of Andalusia, such as Axarquía (Malaga province), the Alpujarra (Granada province), Almanzora and Levante (Almeria province). Increasing attention is being paid to residential tourism in the last of these provinces. Today, many inhabitants of Rodalquilar, a little village nestling in the Sierra de Gata and historically linked to gold mining, are foreigners. As Rodríguez (2015) states, 'they seem to have a good sense of smell for settling in quality-living places (...) and enjoying the fullness of a mini paradise' (p. 47).

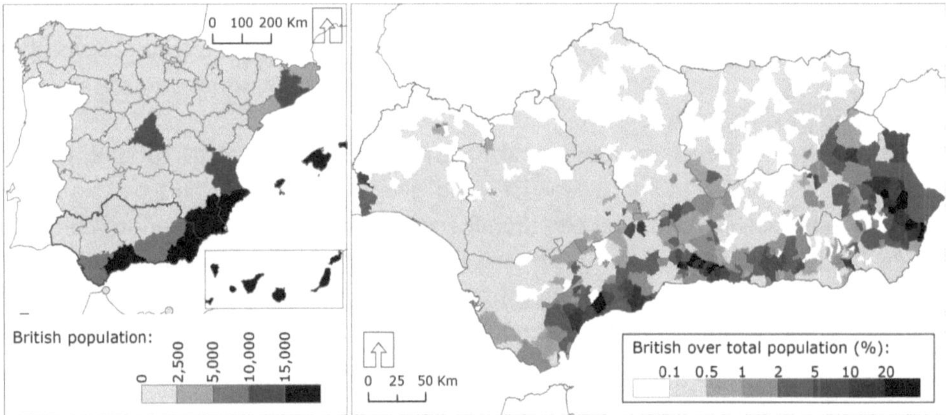

Figure 2. British communities in Spain (provinces) and Andalusia (municipalities). Source: Authors by INE (2018).

According to statistics (own calculations based on INE, 2018), British migration to Andalusia has clearly moved inland. While in 2001 the coastal strip hosted 83.1% of British residents, this figure has decreased to 55.6% in 2017. Inland migration is particularly intense in Almeria province, where today only 29% of British live in coastal municipalities (down from 70.3% in 2001). Thus, geographical dispersion from the coast to the interior has been observed in a short period of time (2001–2017), i.e., the diversification of areas where British immigrants are willing to live (Gualda, 2012).

To better frame the objective of this study, the complementary role played by tourism must be commented on: firstly, because the boundaries between residential and tourism flows are fuzzy (residential tourism, part-time seasonal living, etc.). In fact, today's tourists are tomorrow's potential home-buyers, as many repeat tourists eventually purchase a residential property in Spain or Andalusia; and secondly, because the advertisements analysed below also target tourism investors (Rural Hotels, B&B, etc.).

Spain is the number one destination for UK travellers, accounting for 13 million visits and for 19.8% of the total number of visits abroad in 2015 (Office of National Statistics, 2015). At the same time, the British are the leading foreign tourism market in Spain, with around 25.4% of international arrivals. As illustrated in Figure 3, the number of British visitors has grown steadily since 1999. According to recent information, in spite of Brexit, British tourism is registering record numbers in Spain (Gastesi, 2016). The Canary Islands (26%), the Balearic Islands (22%) and Andalusia (17%) are usually the preferred destinations (INE, 2018).

According to Junta de Andalucía (the Andalusian regional government) (2015), 2.44 million British tourists visited Andalusia in 2015, making the UK the region's main foreign market. This equates to 25.6% of international visitors, while Germany, in second place, stands at only 12%. The weather, monuments and beaches seem to be the main motivations, a similar pattern to North European visitors. However, the main difference lies in accommodation: British tourists often stay in houses, whether their own, rented or family-owned (around 38.5% compared to 55% in Hotels).

Housing investment made by foreigners (and by the British, to be more precise) must be commented on to complete this approach. According to Ministerio de

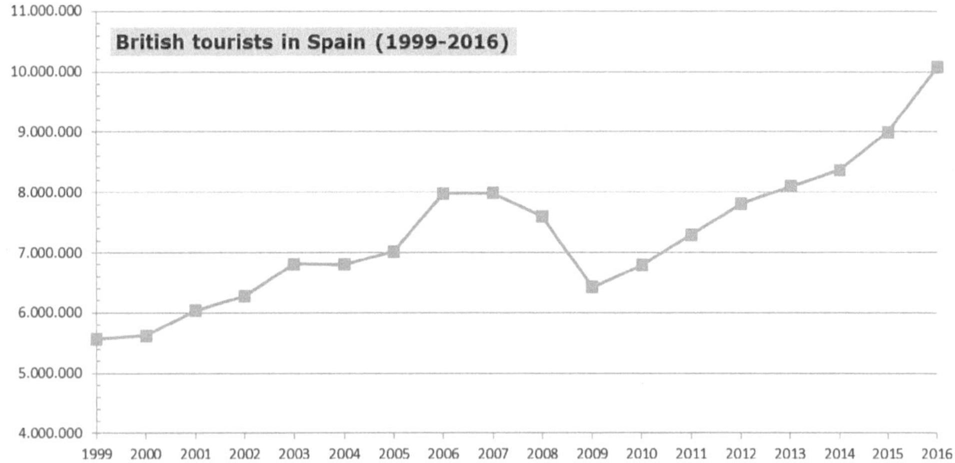

Figure 3. Evolution of British Tourists in Spain (1999–2016). Source: Authors by INE (2018).

Fomento—the Spanish Ministry of Development—(2017), there has been a major rise in real estate inputs by foreigners in Spain since the economic crisis erupted, amounting to 66.4 million Euros during the 2009–2016 period. This increase has been noted in tourism-oriented regions especially, such as the Balearic and Canary Island archipelagos and the Spanish Mediterranean Arc (Valencia, Murcia, Andalusia and Catalonia regions). It is important to highlight the case of Andalusia, which represents a large share of the national figure (22%) and registered a remarkable 124% growth during the same period. Malaga and Almeria provinces stand out from the rest of Andalusia (Figure 4). In the first, real estate transactions by non-resident foreigners accounted for a third (34%) of total housing investment, while in the latter they represent 13%.

The British are at the forefront of housing investment in Andalusia (a quarter of total foreign assets), which might be related to the previously mentioned tourism features. However, while the effects of Brexit appear to have had no impact on tourist data, they appear to have done so on real estate dynamics, as British investment decreased from 21% to 19% of total foreign outlay in Spain in 2016.

4. Methodology

This paper is an empirical approach based on content analysis of residential property development and marketing. It focuses on real estate websites selling properties to British people in Andalusia, particularly those that specialise in rural inland properties. Therefore, two fundamental stages can be distinguished: the first step was to use Google AdWords[1] to analyse the main search concepts related to the purchase of a property and the keywords associated with internet searches of this type. Keyword Planner is a useful tool for determining the average number of searches for any given keywords and their close variants, depending on the target settings and date range selected, and identifying themes and related concepts. Consequently, a range of keywords with similar 'semantic behaviour' can be established. Thus, a specific search, such as *properties for sale in andalucia*, also displays related ideas, such as *house for sale on costa del sol*, *villas andalucia* or *malaga apartments for sale*. The use of

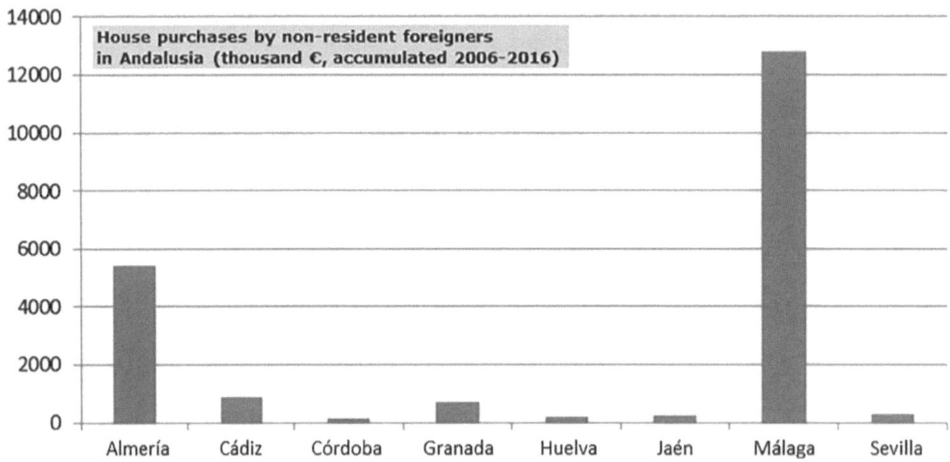

Figure 4. House purchases by non-resident foreigners in Andalusia (Thousand Euros, accumulated 2006–2016). Source: Authors by Ministry of Development (2017).

Keywords Planner to identify sets of concepts with a common semantic behaviour has been successfully applied in several marketing and communication research fields, such as destination marketing and tourism (Xiang, Pan, Law, & Fesenmaier, 2010), environmental marketing (Kronrod, Grinstein, & Wathieu, 2012) and health care and health information management (Ersoy, Ulcay, & Ersoy, 2015; Hartman, 2016; Huesch, Currid-Halkett, & Doctor, 2014).

The second step was to select a set of real estate websites based on the previous main keywords. The sample was drawn from the leading results on two main internet search engines, Google and DuckDuckGo. As the world's most popular search engine, Google dominates the market, with an 82% share in 2017 (NetMarketShare, 2018). Notwithstanding, a complementary search engine was used to compare results. DuckDuckGo is another popular search engine that emphasises the protection of internet users' privacy by preventing tracking. A 'second opinion' was thus ensured when constructing the sample. Despite Google's overwhelming predominance as world leader in Internet-related services and products, their professional ethic has been called into question on more than one occasion, with the European Commission fining the company €4.34 billion for breaching EU antitrust rules (European Commission, 2018).

The sample was filtered and following excluded: (a) very general sites, such as classified advertisement websites operating in multiple countries (Rightmove, Kyero, Zoopla, etc.) and (b) real estate agencies dealing exclusively with coastal areas. Irrelevant or incorrect results were also removed, such as those for the city of Andalusia in Alabama, USA. Once filtered, the sample comprised a total of 27 real estate websites.

Content analysis was used to process the inherent meaning behind the different messages on each website. This technique consists of making replicable and valid inferences from data to their context (Krippendorff, 1980; Weber, 1990). The process of inference underpins content analysis. It seeks to reduce large amounts of information to fewer categories using explicit rules of coding as well as controlled and interpretative deduction (Bardin, 1977). For instance, the semantic concepts of quietness, isolation and peacefulness would be strongly associated with rural areas and white villages, while vibrancy and cheerfulness would be linked to coastal areas. Interpretative codification was alternated with the complementary use of Computer Assisted Qualitative Data Analysis (CAQDAS) to simplify the information into semantic categories.

The aim of this type of technique is to reduce the vast number of semantic elements present in a text—basic units of meaning, such as words (nouns, adjectives), sentences, expressions and other clearly distinguishable textual segments—into fewer, but more complex, semantic cores, such as statements, ideas and concepts. The codification and classification process provides information about the most relevant underlying messages with geographical or territorial meaning (qualitative approach), and allows quantitative comparisons between semantic units.

Nevertheless, some authors have realised that there are some inherent limitations to this method despite its wide application in qualitative studies in the social sciences. As inference is the core intellectual process in this kind of technique, procedures in content analysis have become flexible and, at times, open-ended, with the result that it is considered an analytical instrument without any theoretical ambitions or foundations (Henry & Moscovici, 1968). In addition, as the method is based on text dissection, it sometimes disregards the state of affairs after the texts have been produced, as well as the context that produced them (Titscher, Meyer, Vetter, & Wodak, 2000). This underlying context (motives, attitudes, norms, etc.) is what Henry and Moscovici (1968) call 'conditions of productions of texts'.

Defining the meaning of the basic semantic elements may be another fundamental constraint in this regard. In effect, the determination of meaning must be objective. However, the content analyst has no reliable control mechanisms to guard against subjectivity, since he or she is required to understand the message as conceived by the sender and as interpreted by the receiver.

Hence, content analysis could be considered a description-based, reductive technique. It describes what there is, but might not reveal the underlying purposes of the observed pattern. It is therefore usually applied in combination with other methods. In our case, the foregoing geographical approach (cf. Section 3) is useful for establishing an objective framework that relies on statistical exploration of spatial information sources. As a result, we are able to obtain a clearer and more defined view of the contexts (especially the demographic and socio-economic contexts) in which real estate promotional messages are framed. After all, the use of diverse approaches facilitates the interpretative process that lies at the heart of these techniques by using observed and *objectifiable* facts and figures. A more in-depth understanding of the spatial object of study contributes to balancing the substantial hermeneutical weight involved in the semantic codification and categorization process.

Be that as it may, content analysis has eventually been consolidated as a stable method in social research (Alonso, Volkens, & Gómez, 2012; McMillan, 2000). Over three decades ago Woodrum (1984, p. 1) stated that 'limited application and development of content analysis is due more to unfamiliarity with the method and to its historic isolation from main stream social science than to its inherent limitations'. This type of technique has become commonplace today and has spread to other fields of knowledge beyond its classical applications in anthropology, communication and political science. This widespread use has been achievable thanks to the exponential production and dissemination of texts via the internet in particular.

5. Results: analysis and discussion

5.1. Keywords ideas (Google AdWords keyword planner)

In the first step proposed above, seven seed keywords delivered a total of 615 keywords ideas. Search Network targeting included all languages and search locations in the United Kingdom and a 24 month data range (from July 2014 to June 2016). The following were excluded: excessively vague concepts (house for sale, holidays in spain, etc.), terms with no geographical or semantic link (property for sale in italy, apartments in somerset) and low frequency keyword ideas (fewer than 10 monthly searches on average). Once these corrections had been made, 155 keyword ideas (including seven seed keywords) remained that could be considered a representative example of Google search suggestions based on British Internet users' experiences.

Figure 5 gives a brief classification of the main geographical descriptors and type of housing for the keyword idea sample, weighted according to the total number of searches. As might be expected, the most relevant geographical descriptors or 'place names' were Malaga, Andalusia (and words with the same etymological root, such as andalucia, andalusian), Costa del Sol and Marbella. Regardless of the scale, these geographical scopes were the leading products in both the tourism and real estate sectors. However, apart from these headline results, it is interesting to observe the relative importance of other places located in areas further inland, e.g. Competa, Olvera and Iznajar. A priori, these localities should not be considered major tourism destinations, but Google seems to point in a different direction.

Regarding the type of housing, predictably the largest number of searches included the general concept of property, due to its semantic versatility. However, villa was the second most frequent word used for house, well ahead of all other concepts. The British interpretation of 'villa' is usually a detached house in a rural or suburban environment; the Cambridge Dictionary (2013) defines villa as a 'house, usually in the countryside or near the sea, especially in southern Europe, and often one that people can rent for a holiday'. Villa is the maximum expression of the tourism–residence–southern Europe trinomial, whether on the coast or in rural areas. Despite being less important, *finca* and *cortijo* were also present as genuine forms of Andalusian vernacular architecture among the types of housing keyword ideas.

5.2. Webpage content analysis

The second methodological step involved the analysis of the 27 websites, according to their ranking on two main search engines, Google and DuckDuckGo. The codification of text extracts resulted in a total of 384 basic semantic units, mostly matches between nouns and their corresponding adjectives (stunning and mountains; picturesque, whitewashed and villages, etc.). A set of semantic cores was established from these basic semantic units, as shown in Figures 6 and 7.

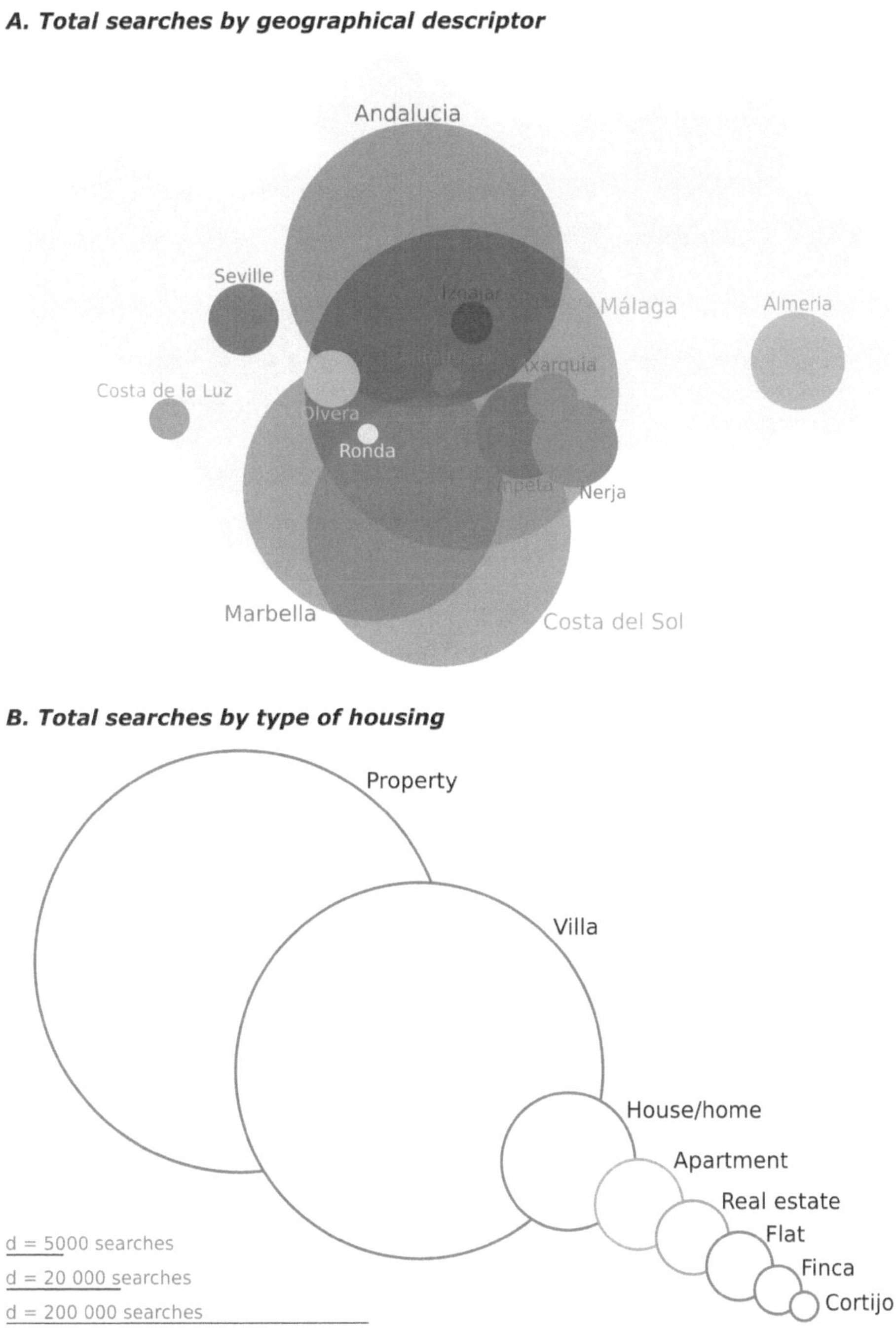

Figure 5. Total searches for keyword ideas classified by geographical descriptor and type of housing (colours are just to make circles easily distinguishable). Source: Authors.

As presented, the Andalusian landscape appears to be mainly configured by natural features and orographic discontinuities (A): mountains, hills, slopes, etc. These are often accompanied by superlative adjectives that reinforce verticality and the rocky, rugged nature of the terrain. Statements are usually overly generic and emphasise the

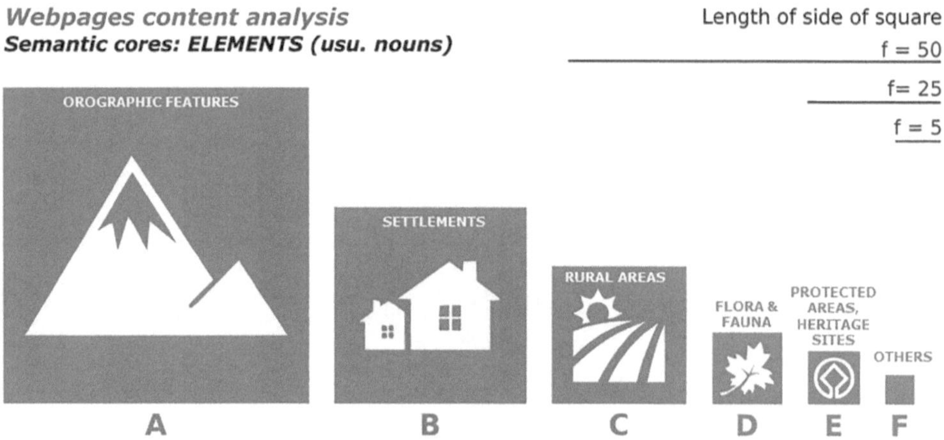

Figure 6. Webpages content analysis: Semantic cores (Elements-Nouns). Source: Authors.

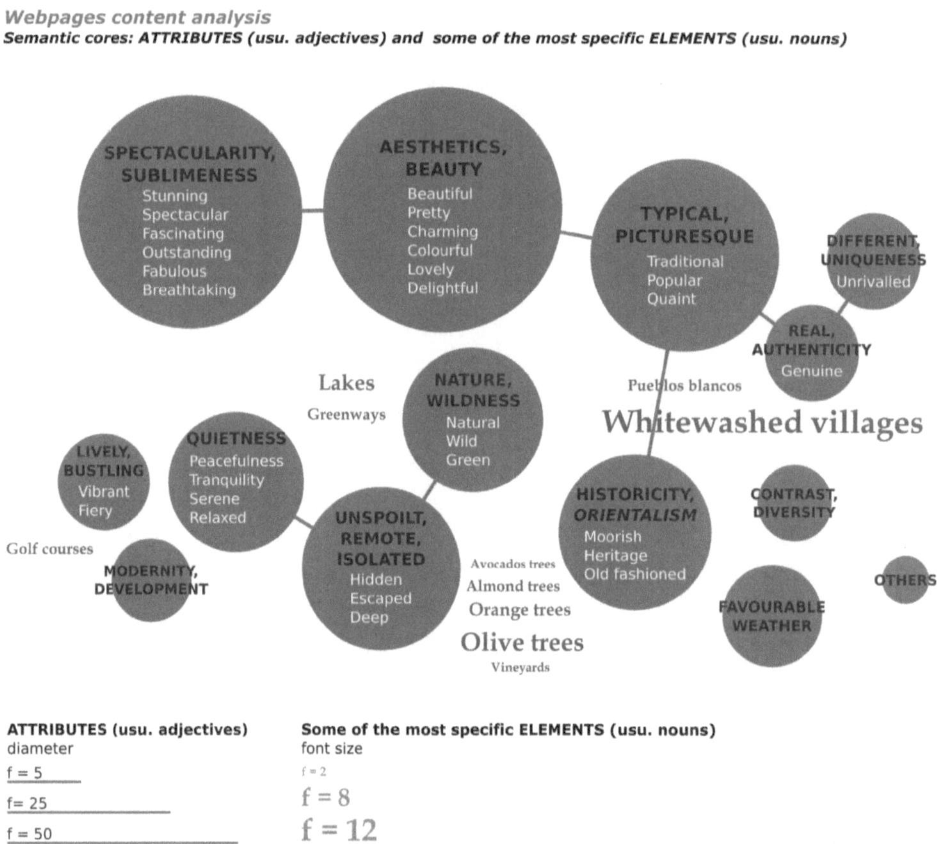

Figure 7. Webpages content analysis: Semantic cores (Attributes-Adjectives). Source: Authors.

value of the views: 'located in the rolling hills of Granada, Montefrío is considered as one of the ten villages with the best views of the world' (Villasfincas). The settlement system is reduced to a set of small villages nestling few and far between in the mountainside (B), with a few details added for good measure, such as meandering streets and whitewashed walls.

The rural landscape is less represented (C) and usually associated with fruit trees, mainly olive groves but also orange, almond and lemon orchards, and other species that the 'British eye' might regard as exotic: 'an area dedicated to olive and almond farming, also orange and lemon orchards and vineyards, a wonderful climate and location for enjoying walking and picnics' (Rio Estates). Barely any mention is made of other traditional agricultural landscapes, such as croplands or grain fields.

There are barely any references to natural ecosystems, flora and fauna (D), and when mentioned, they are linked to the wildness of the region: 'the mountain is very green, the wild flowers grow abundantly and the mountain goats roam freely' (Inland Andalucia). Elements of territorial heritage, such as natural protected areas and cultural heritage sites (E), are rarely mentioned, and the few references that can be found (not necessarily to monuments) usually involve mills, bridges, agricultural techniques and other components of the rural landscape.

The predominating attributes used to qualify the above landscape components (Figure 7) all relate to beauty and aesthetic qualities, especially any adjectives that highlight the sublimeness and spectacular nature of the land: 'It is a typical Andalusian village, nestled in some of the most stunning countryside with spectacular views of the surrounding mountains' (Rio Estates). Other relatively important semantic cores are references to the picturesque and 'typical' natural environment, together with some other related ideas (not necessarily synonyms), such as authenticity and uniqueness. When grouped by semantic proximity, the cores related to isolation and unspoiledness, wildness and quietness stand out. It is interesting to observe that most of the websites contrast the idea of peace and quiet that characterises inland rural Andalusia with the hustle and bustle of coastal areas, especially the Costa del Sol. Nevertheless, this image of a vibrant, festive Andalusia can occasionally be extended to rural Andalusia: 'There are two sides to Spain. There's the cool, trendy and urbane, typical of Barcelona and the Balearics, and there's the wild, fiery side of flamenco dancing and fighting bulls, where the people are fun and noisy and demonstrative, that's Andalucía' (A place in the sun).

There were numerous allusions to a 'rich' historical background, in particular to the legacy of the Moors (despite confusion around periods and cultures), as it is linked to the most distinctive features of the urban settlements (white villages, cobbled streets, etc.), with other decisive historical periods completely ignored. Lastly, a significant number of statements referred to favourable weather (hot summers, mild winters, days of sunshine, etc.), even though this argument has usually been used to promote coastal destinations. Be that as it may, in some cases these descriptions were over-exaggerated and non-rigorous, constituting what might be called climate hype: 'Temperatures that make for (...) gloriously hot summer days' (Country Properties); 'Andalucia has one of the best climates in the world (...). Torrox Costa officially has the best climate in Europe' (Andalucia Casas).

Apart from the direct interpretation of the results via elements and attributes, the analysis also revealed some implicit ideas (ideas that could be 'read between the lines') that ran through the discourse of the different websites:

- In relation to Landscape Character Assessment, which defines the character of a landscape as 'a distinct, recognisable and consistent pattern of elements in the landscape that makes one landscape different from another, rather than better or worse' (Swanwick & Land Use Consultants, 2002, p. 8), landscapes are generally outlined using very generic descriptive formulations. These descriptions are full of common geographical features (mountains, hillsides, valleys, villages, etc.), while more specific features are not well represented. For instance, the elements that form the territorial heritage (both natural and cultural), such as the Torcal karst outcrops or the traditional irrigation systems (*acequias*), are rarely mentioned.
- The various parts of Andalusia are treated unequally. Malaga province and the rest of the Baetic Mountains are much better represented than other areas (the Guadalquivir Valley and the Sierra Morena mountains), due to the higher concentration of British residents, tourists and property buyers there. The non-central areas of the region such as Western Andalusia (Huelva, Seville and Cadiz provinces) and Jaen and Granada provinces hardly appear as descriptors at all, in spite of their relevance in both the region's history and its tourism industry. This unequal treatment could explain, for instance, the almost total lack of references to the dehesa (the Andalusian grasslands), one of the most extensive and characteristic agro-ecosystems in southern Spain; the arable lands, the region's most common agrarian landscape; and the Mediterranean Forest, which represents almost a quarter of all the territory and is the main focus of natural protection policies. Regarding the latter, the under-representation of the Protected Areas Network in the Housing Market's projected image stands out in contrast not only to its spatial occupation (almost a third of the region) but also to its priority role in institutional advertising campaigns. Something similar occurs to a lesser extent with other relevant landscapes, such as sandbank forests, deserts and marshlands.
- These imbalances sometimes result in a distorted construction of the territory. This way of conceiving Andalusia puts Malaga at the heart of the region, and the further from the Costa del Sol, the more peripheral the regions become. The map (Figure 8) taken from Andalucia.com is a perfect illustration. Since proximity to the country of origin an important motivation, Malaga plays a core role as an easy connector with home. The same could apply to older and conditioned British people, especially.
- Despite relevant references to its rural and agricultural landscapes, the region is often depicted as a quasi-wilderness composed of natural, 'untamed' areas. This is partially true, but it cannot be denied that, like many Mediterranean regions, the territory is, above all, a man-made land, fashioned by centuries of human occupation (Braudel, 1989). For instance, it is interesting to observe that the many (man-made) reservoirs and dams in the region are usually regarded as lakes; i.e. natural water bodies (hydrological hype).
- Lastly, the profound contrast between coastal and inland areas observed in a number of regions takes on another perspective in Andalusia. In this case, the dichotomy between the Costa del Sol and the rest of the region (including other coastal areas) is much more evident.

At this point, it is interesting to observe that regional tourism policy and planning have taken a relatively different path from the real estate market in the social

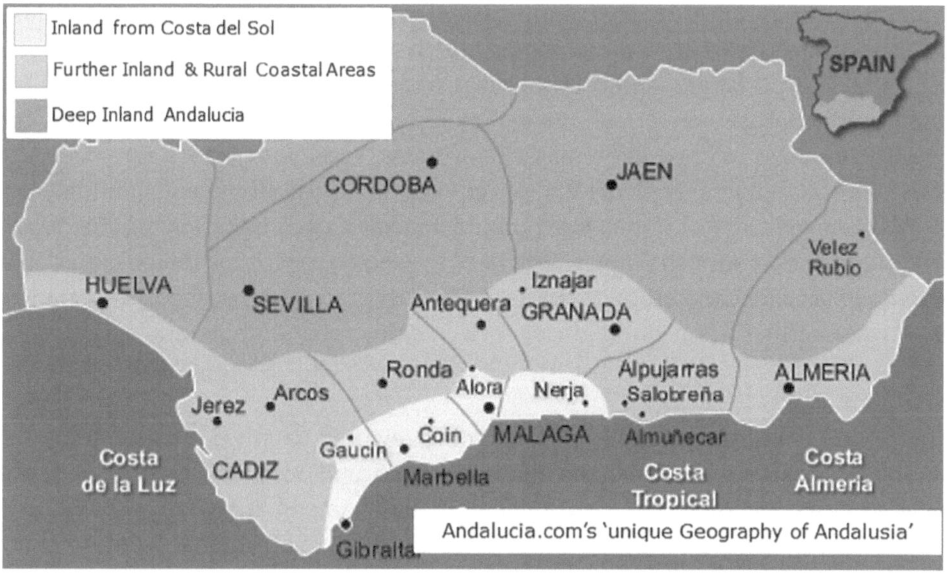

Figure 8. Andalucia.com's 'unique Geography of Andalusia'. Source: Andalucia.com.

construction of a leisure image. In the 1990s, when competences for tourism policy were transferred from the national to the regional administration, the Andalusian government took the lead in this kind of public policy (Merino, Betanzos, & Dorado, 2013). As a consequence, a general image was adopted of the Andalusian region as a destination that embraced its diversity but possessed a prevailing sense of unity, and *Andalucía* was turned into a brand name through a series of promotional campaigns. This reinforcement of the unitary image of the region can be understood as a political entity that legitimises its legal status by asserting a proper identity (Afinoguénova, 2015; Hernández, 2008), particularly to differentiate it from Spain (Martinez & Nicolás, 2014).

Tourism planning in Andalusia has similarly been steered towards a strategic, sustainable vision during the past decade, with greater importance given to the local and sub-regional intervention scales. This has resulted in a geographical broadening and diversification of public investment to improve the tourist offer, either in consolidated coastal and monumental destinations or in those that are emerging inland (Merino et al., 2013).

Thus, the misrepresented image shaped by countries from which customers originate, such as the UK, which is well represented in the case of real estate market, contrasts with the politically unified (but diverse) image of Andalusia and geographical diversification observed in tourism regional planning.

6. Conclusions

At this point, it is appropriate to draw some final considerations to serve as conclusions that might generate some future research lines.

(a) The social practice of producing and reproducing a discourse around Andalusia as a destination is part of a more complex and stepped process where motivations, push factors, expectations and (occasionally) returns are involved (cf. Figure 1). The present paper focuses on a less studied projection of the destination's image; and reinforces the idea of how, as O'Reilly (2014) states, social imaginaries are shaped by actors other than migrants alone, e.g. governments and tourism agents and developers. As observed, in Andalusia, the housing market's deconstruction of the region sometimes contrast with the unitary image of the region, which institutions present as contemporary, differentiated and diverse. Thus, the present paper feeds into a wider line of research that has been productively worked by other authors, and especially, to address one of its main gaps: the housing market as a socially-constructed image actor.

(b) Closely linked to the above, this study has revealed what we consider to be this contribution's main conclusion: the misrepresented and romantic projection of the image of Andalusia still persists to a certain extent. Among the main elements and attributes that characterise the landscape are all those that emphasise the region's sublime, picturesque and exotic nature. Other main landscape qualities and unique areas, specifically, any that today enjoy natural or cultural legal protection due to their heritage value, are less important in this cultural construction.

To a certain degree, this kind of process can be expected of an economic sector such as the housing market, which is more interested in making a profit than in selfless motivation (an informative or educational goal). In these circumstances, the landscape operates as a de facto commodity, as a subsidiary to the main purpose of selling houses and holiday destinations. Thus, the landscape's complexity as a cultural expression is reduced to a set of simplified features that can be easily digested by a social group focused on the materialized experience of a place. In this regard, Aledo (2008) aptly points out that residential tourism contributes to a symbolic and material production of a substitute world, offering a 'hyperreal' experience. The real estate market promises multiple identity qualities to customers through its multiple proposals of consumption. In a globalised context in which local identity is reinforced by global processes, the most competitive elements and features of the landscape prevail, displacing territories and images that are less attractive in terms of economic profitability. In such circumstances, landscape seen as nature or identity is turned into landscape as a product.

(c) From a sociological perspective, the observed romantic projection of the region to captivate the British might cause disenchantment, returns to their country of origin or a re-interpretation of Andalusia. In fact, as Salazar (2014, p. 125) points out, 'one of the central challenges related to imaginaries is the lack of correspondence between the projected ideals and aspirations on the one hand and the perceived and experienced reality on the other'.

In light of all this, this projected image might be expected to evolve and become even more complex and nuanced, as economic, demographic and tourism circumstances change. It will therefore be important to pay attention to the new European framework (Brexit) and its consequences for the real estate market. There are also other circumstances, such as the Costa del Sol being saturated with tourists, emerging inland destinations (linked to more sustainable conceptions), new flight connections, etc., that may affect tourist and residential flows. Consequently, variations in the perceived image of the region might also be noted.

(d) Furthermore, from a territorial scope, the spread of residential tourism is a controversial issue. The image of Andalusia projected by real estate agents could set the construction sector in motion inland, especially in the pre-coastal mountain ranges, by taking advantage of the rural idyll as a marketing strategy and its position as the link between the modern-littoral and traditional-inland Andalusia. In any event, the Regional Administration appears to be aware of the risks attached to this spread (and of a possible second housing bubble), as it has used spatial planning (Plan de Ordenación del Territorio, approved in 2006; Plan de Protección del Corredor del Litoral, approved in 2015) to limit new urban developments, which will inexorably steer housing market towards *second-hand* houses.

Finally, on the basis of the above conclusions, some prospective lines of research can be outlined. On the one hand, the present study was based on the analysis of spoken and written language, and the results might lead to a broader future study that includes other forms of 'language', such as pictures and audiovisual formats. These kinds of communication have become more important due to the contribution that they make to visual culture in an emerging communicative frame based on more than just verbal texts.

In addition, this investigation has highlighted the role of certain inland areas in Andalusia, where relevant emerging destinations for the British exist, and this could also represent a new opportunity of research. Particular attention should be paid to Almeria province, along the southern coast from Malaga, which is a consolidated and recognized destination for British travellers and residential tourists and migrants. The easternmost part of Andalusia. Almeria province has emerged as an alternative location in the quest for new places and lifestyles. What is remarkable about this province is that in recent years inland destinations have become even more relevant than places along the coast.

Lastly, when the analysis of the projected image of the region has concluded, a more in-depth study of perceptions could be conducted with direct exploration among new British visitors. The implementation of qualitative research methodologies (surveys, interviews to stakeholders, focus groups, etc.) will help to strengthen, spread and contrast the main outputs of this contribution.

Disclosure statement

No potential conflict of interest was reported by the authors.

Notes

1. Google AdWords works as an auction system to establish the price of the paid searches shown on any Google search engine results page (SERP). In fact, advertisements have become the main source of income for Google's parent company, Alphabet Inc.: $110.8 billion revenue in 2017, which represents 44% of the global online advertising market (Alphabet Investor Relations, 2018; Smith, 2018).

References

Afinoguénova, E. (2015). Andalucía: una Comunidad Turística sin andaluces. In A. Gómez & J. M. Pino (Eds.), *La retórica del Sur: representaciones y discursos sobre Andalucía en el período democrático* (pp. 229–263). Sevilla: Alfar.

Aledo, A. (2008). De la tierra al suelo: la transformación del paisaje y el nuevo turismo residencial. *Arbor, 184*(729), 99–113.

Alonso, S., Volkens, A., & Gómez, B. (2012). *Análisis de contenido de textos políticos. Un enfoque cuantitativo.* Madrid: Centro de Investigaciones Sociológicas.

Alphabet Investor Relations. (2018). Alphabet announces fourth quarter and fiscal year 2017 results. Retrieved from https://abc.xyz/investor/pdf/2017Q4_alphabet_earnings_release.pdf

Andrews, H. (2011). *The British on holiday. Charter tourism, identity and consumption.* Bristol: Channel View Publications.

Bardin, L. (1977). *L'analyse de contenu.* Paris: Presses Universitaires de France.

Bell, M., & Ward, G. (2000). Comparing temporary mobility with permanent migration. *Tourism Geographies, 2*(1), 97–107.

Benson, M. (2011). *The British in rural France. Lifestyle migration and the ongoing quest for a better way of life.* Manchester: Manchester University Press.

Benson, M., & O'Reilly, K. (2009). Migration and the search for a better way of life: A critical exploration of lifestyle migration. *The Sociological Review, 57*(4), 608–625.

Bertuglia, A., Sayadi, S., López, C. P., & Guarino, A. (2013). El asentamiento de los neorrurales extranjeros en La Alpujarra Granadina: un análisis desde su perspectiva. *Ager, 15*, 39–73.

Betty, C., & Hall, K. (2016). The Mith of No Return? Why retired British migrants in Spain return to UK? In K. Torkington, I. David, & J. Sardinha (Eds.), *Lifestyle migration in practices* (pp. 123–137). Newcastle: Cambridge Schoolars Publishing.

Braudel, F. (1989). *El Mediterráneo. El espacio y la historia.* México: Fondo de Cultura Económica.

Breuer, T. (2005). Retirement migration or rather second-home tourism? German senior citizens on the Canary Islands. *Die Erde, 136*(3), 313–333.

Cáceres-Feria, R., & Ruiz, E. (2017). Forasteros residentes y turismo de base local. Reflexiones desde Alájar (Andalucía, España). *Gazeta de Antropología, 33*(1), a06.

Cambridge Dictionary. (2013). *Cambridge advanced learner's dictionary* (4th ed.). Cambridge: Cambridge University Press.

Casado-Díaz, M. A. (2006). Retiring to Spain: An analysis of differences among North European nationals. *Journal of Ethnic and Migration Studies, 32*(8), 1321–1339.

Casado-Díaz, M. A., Kaiser, C., & Warnes, A. (2004). Northern European retired residents in nine southern European areas: Characteristics, motivations and adjustment. *Ageing & Society, 24*, 353–381.

Cohen, S., Duncan, T., & Thulemark, M. (2015). Lifestyle mobilities: The crossroads of travel, leisure and migration. *Mobilities, 10*(1), 155–172.
Cosgrove, D. E. (1984). *Social formation and symbolic landscape*. London: Croom Helm.
Eimermann, M., Lundmark, M., & Müller, D. K. (2012). Exploring Dutch migration to rural Sweden: International counterurbanisation in the EU. *Tijdschrift Voor Economische en Sociale Geografie, 103*(3), 330–346.
Ersoy, M. A., Ulcay, M., & Ersoy, H. T. (2015). Health information-seeking behaviors about attention deficit hyperactivity disorder on the internet and the analysis of top ranking websites. *Klinik Psikofarmakoloji Bulteni, 25*(2), 178–182.
European Commission (2018, July 18). Antitrust: Commission fines Google €4.34 billion for illegal practices regarding Android mobile devices to strengthen dominance of Google's search engine. *European Commission Press Release*. Retrieved from http://europa.eu/rapid/press-release_IP-18-4581_en.htm
Fabra, L. (2016). *Estadística Registral Inmobiliaria. Anuario 2016*. Madrid: Editorial Colegio de Registradores de la Propiedad, Bienes Muebles y Mercantiles de España.
Fernández, A., & Cruz, E. (2011). Territorio y actividad constructora: del" tsunami" a la crisis. Factores explicativos y propuesta de indicadores a escala municipal en Andalucía. *Boletín de la Asociación de Geógrafos Españoles, 56*, 79–110.
Gastesi, A. (2016, November 1). El turismo británico registra cifras récord en España a pesar del Brexit. *La Vanguardia*. Retrieved from http://www.lavanguardia.com
Giner-Monfort, J., Hall, K., & Betty, C. (2016). Back to Brit: Retired British migrants returning from Spain. *Journal of Ethnic and Migration Studies, 42*(5), 797–815.
Gualda, E. (2012). *Población extranjera en Andalucía y sus redes familiares*. Sevilla: Instituto de Estadística y Cartografía de Andalucía.
Hall, K., & Hardill, I. (2016). Retirement migration, the other story: Caring for frail elderly British citizens in Spain. *Ageing & Society, 36*, 562–585.
Hartman, A. E. (2016). Transparency of hope: Ethical issues in marketing cross border reproductive care: A multi-region content analysis of clinic websites. *Marketing Intelligence & Planning, 34*(7), 943–963.
Henry, P., & Moscovici, S. (1968). Problèmes de l'analyse de contenu. *Langages, 11*, 36–60.
Hernández, J. (2008). *La imagen de Andalucía en el turismo*. Sevilla: Centro de Estudios Andaluces.
Hernández, M., Morales, A., & Sauri, D. (2014). Ornamental plants and the production of nature(s) in the Spanish real estate boom and bust: The case of Alicante. *Urban Geography, 35*(1), 71–85.
Hoggart, K., & Buller, H. (1994). Property agents as gatekeepers in British house purchases in rural France. *Geoforum, 25*(2), 173–187.
Huesch, M. D., Currid-Halkett, E., & Doctor, J. N. (2014). Public hospital quality report awareness: Evidence from National and Californian Internet searches and social media mentions, 2012. *BMJ Open, 4*(3), e004417.
Huete, R., & Mantecón, A. (2012). Residential tourism or lifestyle migration: Social problems linked to the non-definition of the situation. In O. Moufakkir & P. Burns (Eds.), *Controversies in tourism* (pp. 160–173). Croydon: CAB International.
Huete, R., & Mantecón, A. (2013). La migración residencial de noreuropeos en España. *Convergencia Revista de Ciencias Sociales, 61*, 219–245.
Huete, R., Mantecón, A., & Mazón, T. (2008). ¿De qué hablamos cuando hablamos de turismo residencial? *Cuadernos de Turismo, 22*, 101–121.
Instituto Nacional de Estadística [INE]. (2018). Cifras oficiales de población de los municipios españoles: Revisión del Padrón Municipal. Retrieved from https://www.ine.es/dyngs/INEbase/es/operacion.htm?c=Estadistica_C&cid=1254736177011&menu=resultados&idp=1254734710990
Janoschka, M., & Durán, R. (2014). Lifestyle migrants in Spain. Contested realities of political participation. In M. Janoschka & H. Haas (Eds.), *Contested spatialities, lifestyle migration and residential tourism* (pp. 60–73). Abingdon, New York: Routledge.

Junta de Andalucía. (2015). *Principales cifras mercados turísticos extranjeros en Andalucía Año 2015*. Sevilla: Consejería de Turismo y Deporte.

Kellner, D. (2004). Media culture and the triumph of the spectacle. *Razón y Palabra, 39*.

King, R., Warnes, T., & Williams, A. M. (2000). *Sunset lives: British retirement migration to the Mediterranean*. Oxford, New York: Berg.

Krippendorff, K. (1980). *Content analysis: An introduction to its methodology*. Newbury Park, London, New Delhi: Sage.

Kronrod, A., Grinstein, A., & Wathieu, L. (2012). Go green! Should environmental messages be so assertive?. *Journal of Marketing, 76*(1), 95–102.

López-Ontiveros, A. (1988). El paisaje de Andalucía a través de los viajeros románticos: creación y pervivencia del mito andaluz desde una perspectiva geográfica. In J. Gómez & N. Ortega (Eds.), *Viajeros y paisajes* (pp. 31–65). Madrid: Alianza.

Lord, M. (2008). Becoming significant: The appropriation of the French rural space by British migrants. In J. P. Diry (dir.), *Les Étrangers dans les Campagnes* (pp. 249–259). Clermont-Ferrand: Presses Universitaires Blaise Pascale.

Maderuelo, J. (2005). *El paisaje. Génesis de un concepto*. Madrid: Abada.

Mantecón, A. (2017). El turismo residencial no existe. Revisión de un concepto y crítica de su función ideológica. *Cuadernos de Turismo*, 40, 405–422.

Martinez, E., & Nicolás, M. A. (2014). The construction of tourist space by public administration and institutional communication: The image of the brand Andalucía as a tourist destination. *Journal of Promotion Management, 20*(2), 181–199.

Mazón, T. (2006). Inquiring into residential tourism: The Costa Blanca case. *Tourism and Hospitality Planning & Development, 3*(2), 89–97.

Mazón, T., & Aledo, A. (2005). Los límites del turismo residencial: el caso de Torrevieja. *Estudios Turísticos, 165*, 77–95.

McMillan, S. J. (2000). The microscope and the moving target: The challenge of applying content analysis to the world wide web. *Journalism & Mass Communication Quarterly, 77*(1), 80–98.

Membrado, J. C. (2015). Migración residencial y urbanismo expansivo en el mediterráneo español. *Cuadernos de Turismo, 35*, 259–285.

Merino, R., Betanzos, J. M., & Dorado, M. J. (2013). La Estrategia de Turismo Sostenible de Andalucía: Elementos fundamentales en el marco de la planificación turística subregional andaluza. *Revista de Estudios Regionales, 97*, 77–111.

Ministerio de Fomento. (2017). *Estadísticas: Transacciones inmobiliarias (compraventa)*. Retrieved from http://www.fomento.es/MFOM/LANG_CASTELLANO/ATENCION_CIUDADANO/INFORMACION_ESTADISTICA/Vivienda/Estadisticas/TransaInmo/default.htm

Müller, D. K. (2002a). German second homeowners in Sweden. Some remarks on the tourism—migration—Nexus. *Revue Européenne Des Migrations Internationales, 18*(1), 67–86.

Müller, D. K. (2002b). Reinventing the countryside: German second-home owners in Southern Sweden. *Current Issues in Tourism, 5*(5), 426–446.

NetMarketShare. (2018). Market share statistics for internet technologies. Retrieved from https://www.netmarketshare.com

OReilly, K. (2007). Emerging tourism futures: Residential tourism and its implications. In C. Geoffroy & R. Sibley (Eds.), *Going abroad: Travel, tourism, and migration. Cross-cultural perspectives on mobility* (pp.144–157). Newcastle upon Tyne: Cambridge Scholars Publishing.

OReilly, K. (2014). The role of the social imaginary in lifestyle migration: Employing the ontology of practice theory. In M. Benson & N. Osbaldiston (Eds.), *Understanding lifestyle migration* (pp. 211–234). New York: Palgrave Macmillan.

OReilly, K. (2017). The British on the Costa del Sol twenty years on: A story of liquids and sediments. *Nordic Journal on Migration Research, 7*(3), 139–147.

Office of National Statistics. (2015). *Travel trends: 2015*. Retrieved from https://www.ons.gov.uk/peoplepopulationandcommunity/leisureandtourism

Olson, E., & OReilly, K. (2017). North-Europeans in Spain: Practices of community in the context of migration, mobility and transnationalism. *Nordic Journal on Migration Research, 7*(3), 133–138.

O'Reilly, K. (2000). *The British on the Costa del Sol: Transnational identities and local communities.* London, New York: Routledge.
Pack, S. D. (2009). *La invasión pacífica: los turistas y la España de Franco.* Madrid: Turner.
Pillet, F. (2016). Viajeros por los paisajes de España: del siglo XVIII a la actualidad. *Cuadernos de Turismo, 38,* 365–387.
Robertson, I. (2014). Viajes de Ford por España. In F. J. Rodríguez Barberán (Ed.), *Richard Ford: viajes por España, 1830–1833* (pp. 72–111). Madrid: Real Academia de Bellas Artes de San Fernando.
Rodríguez, E. (2015). *Campos de Níjar. Cincuenta años después (1959–2009).* Granada: Ruiz de Aloza.
Rodríguez, V. (2001). Tourism as a recruiting post for retirement migration. *Tourism Geographies, 3*(1), 52–63.
Rodríguez, V., Fernández-Mayoralas, G., Rojo, F., & Abellán, A. (2005). Migración internacional de retirados: los jubilados europeos en Andalucía. In V. Rodríguez, M. A. Casado-Díaz & A. Huber (Eds.), *La migración de europeos retirados en España* (pp. 122–149). Madrid: Consejo Superior de Investigaciones Científicas.
Salazar, N. B. (2014). Migrating imaginaries of a better life… until paradise finds you. In M. Benson & N. Osbaldiston (Eds.), *Understanding lifestyle migration* (pp. 119–138). New York: Palgrave Macmillan.
Shelmerdine, B. (2002). The experiences of British holidaymakers and expatriate residents in pre-civil war Spain. *European History Quarterly, 32*(3), 367–390.
Smith, C. (2018). 55 Amazing Google Advertising Statistics and Facts (January 2018). Retrieved from https://expandedramblings.com/index.php/google-advertising-statistics/
Storm, E. (2013). Una España más española. La influencia del turismo en la imagen nacional. In J. Moreno & Núñez, X. M. (Eds.), *Ser españoles. Imaginarios nacionalistas en el siglo XX* (pp. 530–560). Barcelona: RBA.
Swanwick, C. & Land Use Consultants (2002). *Landscape character assessment: Guidance for England and Scotland.* Cheltenham: The Countryside Agency; Edinburgh: Scottish Natural Heritage.
Titscher, S., Meyer, M., Vetter, E., & Wodak, R. (2000). *Methods of text and discourse analysis.* London: Sage.
United Nations, Department of Economic and Social Affairs. (2015). *Trends in International Migrant Stock: Migrants by destination and origin.* Retrieved from http://www.un.org/en/development/desa/population/migration/data/estimates2/estimates15.shtml
Warnes, A. M., King, R., Williams, A. M., & Patterson, G. (1999). The well-being of British expatriate retirees in southern Europe. *Ageing and Society, 19*(6), 717–740.
Warnes, T. (2009). International retirement migration. In P. Uhlenberg (Ed.), *International handbook of population aging* (pp. 341–364). Dordrecht: Springer Science & Business Media.
Weber, R. P. (1990). *Basic content analysis* (2nd ed.). Newbury Park, London, New Delhi: Sage.
Williams, A. M., & Hall, C. M. (2000). Tourism and migration: New relationships between production and consumption. *Tourism Geographies, 2*(1), 5–27.
Williams, A. M., King, R., & Warnes, T. (1997). A place in the sun: International Retirement Migration from Northern to Southern Europe. *European Urban and Regional Studies, 4*(2), 115–134.
Williams, A. M., King, R., & Warnes, T. (2004). British second homes in Southern Europe: Shifting nodes in the scapes and flows of migration and tourism. In C. M. Hall & D. K. Müller (Eds.), *Tourism, mobility and second homes: Between elite landscape and common ground* (pp. 97–112). Clevedon, Buffalo, Toronto: Channel View Publications.
Woodrum, E. (1984). "Mainstreaming" content analysis in social science: Methodological advantage, obstacles and solutions. *Social Science Research, 13*(1), 1–9.
Xiang, Z., Pan, B., Law, R., & Fesenmaier, D. R. (2010). Assessing the visibility of destination marketing organizations in Google: A case study of convention and visitor bureau websites in the United States. *Journal of Travel & Tourism Marketing, 27*(7), 694–707.

OPEN ACCESS

Topologies of tourism enclaves

Richard Ek and Mekonnen Tesfahuney

ABSTRACT

Research on tourism enclaves has relied mainly on topographical understandings of the phenomenon. The focus has been on the ontic, that which is or exists instead of the relational qualities or properties of tourism enclaves. Topographical conceptions thus tend to simplify enclavic processes and attributes that are much more complex than meets the eye. In this article, we make the case for topological understandings of tourism enclaves, based on a relational ontology, as a complement. We thereby strive to offer more nuanced conceptions of tourism enclaves. We depart from Agamben's political ontology to illustrate our claim. Seen topologically, tourism enclaves are not simply spaces marked-off from the norm, but rather constituents of the norm. Tourism enclaves need to be theorized as 'prototypes' or 'laboratories' of new subjectivities (ways of being, relating, and experiencing the world). The tourist thus emerges as a model figure of biopolitics in the contemporary, the norm rather than the exception. The tourist is not that which is abandoned by the sovereign in the manner of Agamben, but rather a free exilant, a subject that self-willingly chooses abandonment. We deploy topological concepts, like Agamben's the ban, the camp, and state of exception. Such a conception, we argue, widens the ontological register or horizon of tourism theory.

Introduction

> Modern society originally quite closed up, is rapidly restructuring or institutionalizing the rights of outsiders (that is, of individuals not functionally connected to the operation) to look into its diverse aspects. Institutions are fitted with arenas, platforms and chambers for the exclusive use of tourists (MacCannell, 1999, p. 49).

MacCannell recognizes tourism as a key agent of structural differentiation in modernity. Tourism is not incidental but rather central to the modern societal and social order. Today tourism 'constitutes the pinnacle of the capitalist way of life' notes Sloterdijk (2014, p. 195). Touristic ways of being and relating to destinations distil more general ways of being and relating to the world. MacCannell's perceptive observation signals to the need to conceive of and theorize tourism enclaves as 'prototypes' or 'laboratories' of new subjectivities, i.e. as formative of ways of being, relating, and experiencing the world (Löfgren, 1999; Minca, 2009; Simpson, 2016). Read thus, tourism enclaves are anything but exceptions to the norm. In parallel, the tourist can be conceived as a model figure or normative figure of the contemporary instead of the incidental or marginal figure he/she is often made out to be (Tesfahuney, 2016).

Enclaves or enclosures come in many shapes and forms – from the plantation spaces of the 16th century, the enclosures of 17th century England, military barracks, sacred spaces, settler colonies, prisons, hospitals, to gated communities and airports. Conventionally, enclaves have been defined as a territory completely surrounded by another territory in contrast to an exclave, which is a territory belonging to a state but located in and separated by the territory of one or several other states (Berger, 2010). In this sense, tourism enclaves are contemporary territorial expressions of apparently enduring forms of spatial, geopolitical organization (Simpson, 2016). Tourism scholars have recently shown growing interest in these enclavic spaces of tourism. The theories, concepts, and vocabulary familiar from political geography – territorialization, bordering, and of course the concept of enclave itself – have been mobilized and applied to shed new lights on this particular touristic spatial phenomenon (Saarinen, 2017). Research on the geopolitical dimensions of tourism as well as 'geopoliticizing tourism geographies' (Mostafanezhad, 2018) is a case in point. Saarinen (2017) provides a conceptualization that captures the dialectic of inclusion and exclusion underlying enclavic tourism. Conceived thus, enclavic tourism spaces are an expression of human territoriality and the regulation of flows, mobilities, resources, and people. One early example is the strand of research in tourism studies that highlighted the backpacker phenomenon and how backpackers engendered tourism enclaves in the name of more virtuous tourism practices and ideologies (Elsrud, 2004; Lloyd, 2006).

However, research on enclaves in tourism studies, has by and large relied on topographical approaches to and understandings of the phenomenon. Studies of enclavic tourism have mainly been occupied with mapping out and examining the ontic dimensions of enclavic tourism, i.e. that which is or exists, instead of the relational qualities or properties underlying tourism enclaves. Geographically, the major thrust of the research has focused on empirical studies of constructed (mostly by private stakeholders) tourism enclaves in the South (Klein, 2007; Saarinen, 2017). Tourism enclaves have been theorized as geographically distinct areas vis-á-vis the surrounding territory, society, culture, and economy – thus more or less detached from the regional economy, polity, and culture they are located in or surrounded by. The focus on enclaves in the South inadvertently bolstered conceptions of enclaves as idiosyncratic spaces, 'aberrations' from normative tourist spaces, ways of being and relating to the world.

Topographical conceptions of tourism enclaves tend to simplify enclavic processes and attributes. Questions of biopolitics, power, and normative subjectivities constituted in and encapsulated by enclavic tourism are lost in topographical renditions as they are not visible from a topographical perspective or through a topographic lens. In this article, we propose a topological understanding of tourism enclaves based on a relational ontology. Topological/relational conceptions of tourism enclaves offer alternative analyses of biopolitics, power, and normative subjectivities constituted in and encapsulated by enclavic tourism. A key implication of the perspectival shift we propose here is that seen topologically tourism enclaves are much more than just spaces that are territorially and functionally marked-off from the norm. Indeed, from the perspectives of topological space, these tourism enclaves are constitutive of normative tourism spaces. The tourist thus emerges as a model figure of biopolitics and geopolitics in the contemporary, the norm rather than the exception. Relying on Agamben's (1998) political ontology to illustrate our claim, we deploy topological concepts, such as *the polis, the ban, the camp,* and *state of exception*. Such a conception, we argue, widens the ontological register or horizon of tourism theory as it complements the topographical view on tourism enclaves predominant in tourism studies.

By adopting relational (topological) perspectives that place the exception at the center of western political ontology, we strive to align enclavic tourism research with wider contemporary philosophical currents. A relational ontology of enclavic tourism questions conventional distinctions between enclavic and non-enclavic tourism spaces. Our readings and topological understandings of enclavic tourism, therefore, transcend conventional topographic (dichotomous) distinctions such as tourist or citizen, inside or outside, inclusion or exclusion, norm, or exception (Ek & Tesfahuney, 2016). Our reading invokes a geopolitical–philosophical reasoning based on Giorgio Agamben's (1998) topological readings of Western political ontology that opens up possibilities for alternative theorizations of enclavic tourism. Consequently, our conceptualization is aligned to the contemporaneous biopolitical 'turn' in tourism studies (Roelofsen & Minca, 2018; Rutherford, 2011; Simpson, 2016) as well as the budding geopolitical turn in tourism geographies in general (Mostafanezhad, 2018).

Our attempts to apply the insights of political philosophy to issues in tourism theory are invitations for further deliberations, dialogues, and debates in tourism studies, and thereby (re)envisage and (re)cast matters touristic anew. In short, this text is a plea to ask new research questions regarding (enclavic) tourism, in which the enclaves and the enclave tourists is conceptualized as the role model of the contemporary world order, whatever it is labelled as neoliberal, post-political, or simply capitalistic. The perspective shift we are proposing also disrupts the scale of continuum between the tourist and the citizen, including 'middle forms' like residential tourism and lifestyle migration. Widening the ontological horizon of tourism studies is thus a means to an end, rather than an end in itself. A widening of the ontological horizon creates theoretical opportunities or possibilities that enrich empirical understanding of not only tourism enclaves specifically, but tourism as a way to relating to the world in general as well.

In order to theoretically and analytically realize the aim of this article, we lay out our argumentation in three stages. *First*, we provide an overview of existing research on tourism enclaves. Simultaneously we outline in more detail the ontological distinctions between the habitual or conventional topographical understanding and topological conceptions of the spatial. We juxtapose the spatial ontology underlying research on enclavic tourism to the political-ontological approach we propose here. The alternative reading provided here is in no way exclusive, and only one of many ways of imagining and examining enclavic tourism in particular and the world of tourism at large in general. *Second*, we introduce and deploy Agamben's (1998) conceptions of the city and the camp as instances or figurations of topological space, stressing the dialectical relations rather than distinctions or separations of dyads such as exception/norm, inside/outside, and citizen/tourist. Thus, we complement conventional notions of tourism enclaves that follow a topographical ontology, whereby the enclave is primarily seen as the exception from the rule. This is particularly prominent in critical, particularly Marxist perspectives. The enclave is conceived as an aberration from the normative spatialities and temporalities of tourism. Rather, we propose that the enclave is more than an idiosyncratic tourism space, but rather the very essence of or distillation of tourism. *Third*, in the vein of Franklin (2004) among others, the previous analytical operations provide the ground for adopting compound ontologies of tourism, yet from a more political–philosophical perspective. Finally, we close this article by proposing new directions and research questions that have emerged from our topological readings of enclavic tourism.

In what follows, in the next two sections, we introduce topographical and topological conceptions of tourism spaces, respectively. We rely on prior research on tourism enclaves to (1) highlight the spatial ontologies underpinning enclavic tourism research and (2) to juxtapose an alternative, topological (relational) perspective on enclavic tourism. We would like to stress that our primary aim here is not to critique prior research, rather the aim is pedagogical. In the rest of the article, we deploy Agamben, as an illustration of how to work with metaphysical and politico-philosophical notions such as state of exception, inclusion/exclusion, the threshold, abandonment/the ban, the camp/biopolitics and power in relation to tourism theory – a very modest attempt to broaden the ontological horizon of tourism research.

Topographies of tourism enclaves

Predominantly, research on tourism enclaves that apply a topographical approach rely on Euclidean geometry and absolute notions of space in the understandings of enclavic tourism. Space is understood as *res extensa*, physical expanse and distance (á la Descartes). Moreover, space is given and stable, and exterior to the events and phenomena it frames (see Massey, 2005; Harvey, 2006). Space is thus conceived as a container. A topographical understanding stresses the (absolute) physical and functional geographical reliefs and distinctions, similarities, differences, and patterns in a three-dimensional Euclidean space. The topographical map is the most typical representation of this particular ontology, one that runs deep in Wester metaphysics (Casey, 2002). The inside/outside distinction is an instance of topographical ontology (Lefebvre, 1991). According to Latour (1993), the creation of these two distinct ontological dichotomous realms or domains has to do more with exclusionary categories and practices of purification that are inherent to and definitional of the modernistic (epistemological) project than with the physical world *per se*. It is an *either/or* rather than a *both/and* logic or principle that is constitutive of modernist, topographic ontologies and generically hierarchical modes of thinking (Haraway, 1991). *Ipso facto*, it is impossible (schematically and figuratively) to be both inside *and* outside, to straddle two different realms at the same time. The distinction, *per se*, is the element in the ontology that creates the difference; the boundary (Nail, 2016), the barbed wire that constitutes the social (Netz, 2004), the walls that created Rome (Diken & Laustsen, 2006). This distinction is an act of (sovereign) boundary making, and as such a foundational partition that literally conjures up Leviathan, in other words the state of nature, and the social contract. Barring this division or distinction, neither enclavic tourism nor tourism spaces for that matter either.

Initially, research on tourism enclaves relied on absolute conceptions of space and its theorizations relied on topographical distinctions and perspectives (Kermath & Thomas, 1992). Later on, the focus of touristic enclave research on destinations in the global south placed the subject in a broader North/South spatial divide and international economic order, as well as in a regional and local context. There was for sure an awareness of the interactions and exchanges between the formal economy of tourism enclaves and the informal economy of the surrounding territory (Kermath & Thomas, 1992). Yet as noted by Freitag (1994), the very spatial design of the enclave made it difficult for the regional industry to get 'into' the economy of the tourism enclave, save the local elite. Interaction between the economy of the enclave and the surrounding economy is very limited. If and when local stakeholders manage to get 'in', they are forced to the margins of the enclave economy (Shaw & Shaw, 1999). In as much as enclavic tourism 'is a form of tourism characterized by physically, socially and economically self-contained structures segregated from the local community' (Naidoo & Sharpley, 2016, p. 18), the socioeconomic impact of the tourism enclave on the surrounding and wider neighboring society has been limited (Anderson, 2011; Mbaiwa, 2005).

Tourism enclaves seen topologically

In a topological understanding on the other hand, non-metric properties of spatial figurations such as connectedness and density are stressed, which implies a focus on functionality, relations, the interactions between relations and, importantly, the processes of spatial emergence below the (topographical) surface (Murdoch, 2006). The topological imagination harmonizes with a relational notion of space. Relational takes on spatiality imply a paradigmatic departure from the Cartesian absolute understanding of space since these dissolves the boundaries and borders between objects and space. Space *is* the product of processes and events rather than a container of processes or an arena where events take place *in* (Smith, 2003). As space is a process, space is also always already in flux and always in emergence. Relational space is based on a Heraclitan ontology of *becoming* while absolute space is based on a Parmenidean ontology of *being*. In a way, relational space is a verb, both socially produced and socially constructed, while absolute space is a noun (Doel, 2000).

Early studies of backpacker tourism, occupy a middle ground between purely topographic approaches to enclave tourism on one hand and the topological conceptions that is on offer here (Cohen, 2006). Although these studies strove to broaden the concept of the enclave, they nevertheless departed from a topographical understanding of space. As the backpacker is, supposedly, a tourist that seeks out the local other, the boundaries between the backpackers and locals are porous. The interaction between the tourist and the citizen of the destination becomes a relative space of dynamic societal interaction (Cohen, 2006; Lloyd, 2006). The backpacker tourism enclave is characterized as a space of social negotiation between hosts and guests, a space of suspension (Wilson & Richards, 2008). Yet and for all that, the enclave is still a space ontologically given and possible to pin-down, map out, and categorize. The typologies of tourism enclaves (Cohen & Neal, 2012) is based on the permeability of the border that separates the enclave from the wider (absolute) space the enclave is placed or situated 'in'. As spelled out by Cohen and Neal (2012, p. 573): 'researchers have noted differences in the relative tightness versus looseness of enclave boundaries: resort enclaves are reported to be tightly bound, while backpacker and urban tourism enclaves are relatively fuzzily bounded and easily penetrable'. The spatial distinction that defines the tourism enclave is still in operation, even though its specific nature varies between different enclaves. From an ontic view, enclaves come in many forms.

Relational understandings of space have been established in tourism studies in general for decades now, and have also made their mark in enclavic tourism research. A relatively early example is the study by Torres and Momsen (2005) in their discussion of Cancun as a socially constructed 'Gringolandias' and their analysis of Cancun as a transnational relational space infused not only by capitalistic power-relations but resistance as well. The resort is a product of the interplay of transnational capital and consumption- as well as production-led migratory flows to Cancun. The intellectual ammunition of the relational space approach comes to good use when the tourism enclave is conceptualized as a socio-spatial process with inherent paradoxical spaces and representations of space as visualizations of power (Carlisle & Jones, 2012). Boundary-work and counter representations confirm Massey's and Lefebvre's contentions that space is always oozing, immanent, in becoming and hence malleable (Buzinde & Manuel-Navarrete, 2013; Manuel-Navarrete, 2016). Arguably, a topological approach to tourism enclaves relies on ontogenesis (immanence and becoming), rather than ontic (transcendence and being) conceptions of space (Nancy, 2000).

In our effort to set new light on tourism enclaves, we pose new questions on enclavic tourism from the perspective of ontogenesis rather than conventional ontic approaches. From the perspectives of ontogenesis, it is possible to be *both* inside *and* outside at the same time as embedded in a topological spatiality. It is a 'both-and – situation' that applies. Think about the Möbius band, a common topological figure: there is no distinction between the inside and the outside of the band as one can move along the whole strip without transgressing any border or threshold. The vey meaning and relevance of the inside/outside – dichotomy comes into question. Topographical attributes like physical distance and relief take a secondary role, as topology primarily focuses on other attributes such as reach, folding and becoming (Allen, 2016). In human geography, the topological views of space have been influenced by French Neo-Marxist theory, French post-structuralism and geophilosophy as well as German existentialism. Shields (2013) 'cultural topology' and the 'post-mathematical topology' outlined by Martin and Secor (2014) are two examples. However, the 'topological turn' is still very much exploratory in nature and more of an ongoing dialogue rather than a full-blown paradigm shift. Moreover, there are philosophical and analytical reservations and uncertainties with regard to the applicability of mathematical topology to the social sciences in general (Abrahamsson, 2012; Barba Lata & Minca, 2016). Perhaps the distinction between the topographical and the topological is altogether a false one? Is it not a distinction made only possible to see from a specific perspective, from a topographical perspective which imply that there exists a correct representation of the world? We argue that it is more productive to imagine or theorize space as an arena of being *and* becoming, immanence *and* transcendence, ontogenesis *and* the ontic, articulated as *both* topography *and* topology. We can draw parallels to David Harvey's (2006) answer to the 'eternal' question of whether space is absolute, relative, or relational. In the end, Harvey settled for space being all three. Is space topographical or topological? We conceive of space as both topographical and topological (Nancy, 2000). Hence, our take on tourism enclaves is complementary to rather than a dismissive critique of topographical perspectives on tourism enclaves.

City and camp

In this section, we present an overview of Giorgio Agamben's understanding of the spatial relation of the city and the camp. Agamben has had a tremendous influence in human geography, not least via the early works of Minca (2005, 2006, 2007), Ek (2006) and Gregory (2004, 2006). Minca has shown how Agamben's political philosophy works on an explicit spatial register. In what follows, we focus on the city and the camp (Diken & Laustsen, 2006; Giaccaria & Minca, 2011). His conception of topological space is also a critique of absolute notions of space inherent in western metaphysics and political philosophy. In Agamben's reading, Western metaphysics has the topographic circle as its foundational figure and model (Ek, 2006). Agamben's philosophy provides an entry point to theoretically flesh-out our arguments and enables us to problematize and show the limitations of predominant ontological conceptions of tourism enclaves.

Sovereign power, as per Agamben, is about making *and* breaking ontological distinctions. Indeed, the sovereign is the one that makes *and* breaks the law – the two primary, or original, acts of sovereign power. The sovereign is located at the threshold of the inside and outside, law and state of exception, life and death, *zoë* and *bio*. The inside/outside ontology is central to Aristotle's notion of the *polis* as the highest good and the community of 'Man' as a political animal (*bios*) in contrast to *zoë*, the bare or simple life of being, or natural life (Agamben, 1998). One of the most famous foundational urban myths, the story of Romulus and Remus, captures as well the dialectics of *zoë/bio*, inside/outside, civilization/barbarism, central in Agamben's deliberations on the city and the camp (Diken & Laustsen, 2006). Sloterdijk (2016) provides a similar conclusion albeit one based on an existential rather than political reading of the dyad civilization/barbarism. Indeed, a similar thought model is at play in Hobbes' Leviathan and his distinction between the state of nature and a commonwealth (Diken & Laustsen, 2005). The spatial corollaries of the *inside/outside, inclusion/exclusion, zoë/bios* distinctions, chisel out territories as absolute spaces and nation states as institutionalized containers of society (Agnew, 1994; Elden, 2009; Häkli, 2001; Norris, 2005; Sassen, 2006; Wall, 2005).

According to Agamben, these western conceptions of the world rest on an illusory distinction. The ontological security promised by the social contract, the *polis* (the political community or nation-state), is always already provisional, as sovereign power has the capacity or power to un-make the distinction, and institute a ban on its subjects – to abandon its subjects (Wall, 2005). Ontically speaking 'we' may feel safe in Leviathan – an apparently benevolent sovereign that protects us as subjects, yet from an ontogenesis point of view 'we' are never safe even on the 'inside'. Rather, we occupy the space of *indistinction*, at the threshold of *zoë/bio* (Gregory, 2004). Thus, the topographical inside/outside model fails to fully capture the power-topology underlying sovereign power (Agamben, 1998). As per Agamben, the camp is the exemplary space where sovereign power is spatially played out. The space of the camp is ontologically speaking a threshold in constant transition, with subjects neither in nor out but caught betwixt (Koopman, 2008). Physically, camps are spatial manifestations of this sovereign power and may appear ontically in malevolent (prisons, concentrations camps) or benevolent guises (airports, shopping malls, event arenas, tourism enclaves). One should thus speak of camps (in the plural) since as Agamben argues, camps have a paradigmatic position in the contemporary biopolitical order and come in many shapes and ways. They range from torture camps/prison spaces such as Guantanamo and Abu Ghraib, to all-inclusive resorts, ludic leisure spaces, and hedonistic/eudainomic experiencescapes. Still, they all have the same underlying spatial logics of *zoë/bio, inside/outside, inclusion/exclusion.* In brief, they are articulations of sovereign power and control, be it economic, political, cultural or social, and not the least spatial (Diken & Laustsen, 2005; Norris, 2005), and eventually articulations of *life/death* (Foucault, 2003, p. 240–241).

Tourism enclave – now you see it, now you do not

On one hand, a tourism enclave is 'simply' an ontic actualization, a material manifestation absolutely worthy of careful mapping and investigation. At the same time one that does not tell us the whole story. The logic of the 'enclave', like the logic of the camp, requires musing and interpretations that transcend topographical categories and boundary markers. In the topographic frame, the tourism enclave is a conceptual and spatial construct that gives priority to its visuality, i.e. the topographic frame is *ocular-centric*. This is no coincidence as a topographical view of the world comes 'naturally' due to the hegemony of vision (Rodaway, 1994). A scopic regime with perspectivism as a dominant visual practice (Jay, 1988; Levin, 1993) lends itself to an inherently topographical view of the organization of society and the ordering of territories. It is through a topographical view that walls, borders and entry points become visible and distinct in the first place. To David Michael Levin this means that the world view that the visual practice makes possible is 'frontal': that is 'ontology of entities which, at least in the ideal situation, are to be held 'front and centre': in the most ideal act of *beholding*, the object is to be *held* in place directly before the eyes. What is laid out here is a metaphysics of vision, one that tends to overvalue constancy, uniformity, permanence, unity, totality, and distinctness' (Levin, 1989, p. 31, e.o.), that is, a world-picture (Heidegger, 1977; Gregory, 1994). The tourism enclave is a part and parcel of the world-picture and the world as exhibition (Mitchell, 1989).

However, a far-reaching understanding of enclavic spaces in tourism needs to combine topographic and topologic lenses in analyzing processes of enclavization, involving both the opening-up and closure of space under the sign of tourism. What is not immediately displayed in the tourist world picture, is the commodity relation that upholds the enclave. Just like the commodity, capitalist relations between people are hidden behind exchange relations in the market place. The social organization that constitutes the tourism enclave is mediated through the market place. The subjective and abstract values of the tourism enclave are reduced to objective, material things (Marx, 2002). The touristic enclave is 'only' the materialization and empirical manifestation of a phenomenon that, just like the commodity, externalizes or bleaches out the underlying capitalist relations. Viewed topologically, the tourism enclave is a materialization or concentration of capitalistic relations, social order, and practices, that are actually inherent to society at large (Büscher & Fletcher, 2017). The enclave is not an exception, rather (part of) the norm. As an 'empirical beacon', the tourism enclave may easily give the impression that tourism is spatially (de)limited and clearly demarcated to certain physical areas. Moreover, you are a tourist only when certain conditions are upheld, like an overnight visit *in* another (topographical) place. This topographical imagination of tourism as something outside of or marginal to society at large places greater value on research questions that approach tourism as contained and demarcated in time and space. However, if the topographical imagination is supplemented by a topological one, tourism enclaves emerge as something more than enclosures, or carved-off idiosyncratic spaces. For instance, in a topological imagination the relationship between the tourist in the enclave and the citizen outside of the enclave become intertwined. Where tourism or the enclave ends and where the polity or citizen begins is not clear cut. In a topological imagination, both the tourist-citizen and the enclave-polity lie at the threshold, i.e. the spaces of indistinction. A major implication of this is that the tourist and the citizen are but two faces of a topological figure or relation. A stance that may set the sirens of tourism scholars on, since if this is the case then everyone may be seen as a tourist. Alternatively, no one may be seen as a tourist. However, this is not what we are implying. Instead, in a similar vein to Foucault's conceptualization of power, we uphold that tourism consists above all of a set of practices, dispositions, ways of being in and relating to the world that permeate the contemporary societal.

The citizen and the tourist

The topological reasoning outlined above with the Möbius strip as its frame or model rather than the circle of the topographical (with equidistance, clearly marked inside and outside, inclusion and exclusion, bordering and ordering, territorialization, and cartographies of space as fixed and demarcated), offers new ways of seeing and thinking tourism enclaves. By this we mean novel ways of contextualizing, theorizing and describing the phenomenon of tourism enclaves in the field of tourism studies that in particular focus on its manifold spatialities. The specific reasoning applied here can be 'up-scaled' and extended to shed light on established conceptions of tourists, tourism and thereby the ontologies and epistemologies of tourism theory as well. Indeed, one could argue that tourism is always already enclavic: it is apart from everyday life, the authentic and extraordinary, a space where quotidian norms, values and relations are spatiotemporally 'suspended' so to speak. Tourism studies, its canon, has precisely this as its point of departure.

Tourism as a phenomenon has conventionally been conceived, based on a topographical ontology, on a foundational idea in which tourism is viewed as something different, separated out from, or beyond quotidian life. As such, outside of work, family life, politics and the spaces of dreary everyday existence, a logic of particularity, of the unique, of the non-quotidian, underlies the canon and foundational ideas of tourism. A logic of particularity that always already is inherent to the discipline of tourism studies and permeate its world-view. The world as an oyster of extraordinary, distinguished, and exceptional touristic places for the (usually male) tourism scholar to explore and conquer (Ek & Larson, 2017), in stark contrast to the mundane and ordinary home (Nilsson & Tesfahuney, 2018). Tourism has been framed as the search for and the experience of the authentic other (so central to the discipline's core conceptual apparatus) and the spatiotemporal sublimity of mythic and extraordinary touristic places (nowadays increasingly transformed into alienating touristic enclave by a relentless market-based commodification machinery), or if you like, touristic spaces of exception. But again, these touristic spaces of exception – tourism enclaves spread all over the world – from a topological viewpoint – are not ontologically exceptional, rather 'only' actualizations of force-fields, of near and far, of here and there, of subject and sovereign, of law and state of exception.

By rethinking institutionalized concepts like 'tourism' and 'tourist' through less institutionalized concepts like 'tourism enclave' it becomes possible to reflect upon tourism as a societal phenomenon in new ways. This has been done earlier in the work of Franklin (2004, 2007) and Edensor (2000, 2006) among others, but while Franklin discusses tourism theory through the perspective of social theory and Edensor from a practice and performativity perspective, we apply political philosophy to arrive at our conclusions. From a political-philosophy perspective, it is particularly expedient to question and investigate the binary pair of the 'citizen' and the 'tourist'. The tourist is by tradition understood and juxtaposed as someone distinct or different from the political citizen, the member of the *polis*, the community and as such mapped into the inside/outside – distinction. As such the tourist is rather akin to the stranger, the outsider, the alien. 'Proper' social science disciplines like sociology and political science as well as disciplines like philosophy, human geography and history have been highly attentive to the figure of the citizen, in stark contrast to the disdainful views of the tourist and the field of tourism studies (Nietzsche's [1996] remark about stupid and perspiring tourists climbing mountains speaks for many). In tourism studies, the famous host – guest model is built on this duality and even if we have several scholarly works that problematize this distinction (Rojek, 1998), or unfold the disentanglements between the tourist and the citizen (Bianchi & Stephenson, 2014), by and large the distinction between citizens and tourists is still taken for granted in tourism studies. The current debate on over-tourism is one example. The condescending view of the social sciences to matters touristic is but a species of topographical logic, which when viewed topologically reveals that the tourist has more in common with the citizen proper – something that even the celebrated thinkers in philosophy (including, perhaps ironically, Agamben) and other scholars of the citizen fails to espy.

However, from a topological perspective, the tourist and the citizen positioned side by side is a picture that does not hold in the same way as the distinction between the tourism enclave and whatever lies around the enclave does not hold. Rather, the relationship between the political citizen and the non-political or perhaps better post-political tourist is better depicted as a Möbius strip: the tourist is the flipside of the citizen, the citizen the flipside of the tourist. This implies that the tourist is perhaps an empirical phenomenon that can be dated to the industrial era or the post-industrial era, as commonly understood, but actually has an older philosophical ancestry, as the ontological *doppelganger* and constant/shadow companion of the citizen. Similarly, the camp has been theorized by Agamben as an inherent principal order of the city and not, as Foucault argues, a colonial invention (Agamben, 1998). Be that as it may, neither are the silent companions that they once presumably were. Agamben provocatively argues that the logic of the camp is nowadays generalized – it is the paradigm of social life and subsumes the whole society (Diken & Laustsen, 2005). The same trajectory may be traced for the tourist as a paradigmatic figure – the mobile sovereign tourist subject, one who decides the exception and establishes normative mobile subjectivities, ways of being in and relating to the world. In short, the tourist as the ideal typical subject of our time (Tesfahuney, 2016). The injunction *Enjoy!* that defines contemporary life (Žižek, 1996) is but a generalization of the touristic way of life. The tourist subject is thus not marginal to, but integral to everyday (social) life (Edensor, 2007; Franklin, 2003; Hannam, 2008; Pernecky, 2010), in which the development and growth of tourism (enclaves) is but one instance or empirical indication of (Sidaway, 2007). Moreover, the tourist is anything but the a-political subject that he/she has traditionally been predominantly theorized as. Rather, a political subject *par excellence*, if only as an agent of de-politicization (hedonism, enjoyment) or side-lining of politics proper (Kingsbury, 2005). A topological imagination thus brings forth the problematics of contemporary theorizations of the post-political, by Slavoj Žižek (1999), Jacques Renciére (1999), Chantal Mouffe (2005), Mustafa Dikeç (2015) and Erik Swyngedouw (2018), that fail to recognize and give proper due to the tourist subject in their mappings of the post-political contemporary.

Coda

> [B]iopolitics begins with enclosure-building (Sloterdijk, 2014, p. 170).

To recapitulate our argument, we propose a topological understanding of tourism enclaves that go beyond conventional topographic takes on tourism enclaves as distinct geographical articulations of tourism. We argue that the logic of the tourism enclave permeates much of what at face value is not considered enclavic tourism. Thus, it could and should be searched for elsewhere, particularly in places where it is least associated with or expected, as the privatized public library or in the uncanny home (Nilsson & Tesfahuney, 2018) and not only where it manifests itself empirically. That said, we do not claim that the world is literally filled with tourism enclaves, rather the logic of the enclave is much more complex and hence tourism enclaves need to be 'gazed anew'. The question is how. This article has attempted to provide new ways of looking and studying that.

To 'see' or 'apprehend' the logic of the enclave, tourism theory needs to develop topological eyes that discern the distant as proximal, the exception as the norm and new cartographies that rupture its habitual geographical imaginations and ontological standpoints. Political theories and philosophies that rupture by invoking concepts that cipher the world topologically, such as Agamben's politico-philosophical vocabulary, provide such a vision and a cartography of enclavic tourism. In what follows, we outline some of the directions that tourism geographies could fruitfully take.

A first set of questions relate to further inquiries to tourism spatiality for instance along the lines of Cantillon (2018). Thus, while the spatiality of the camp in the ontic shape of asylum détente centers has been mapped out extensively, the spatiality of the tourism enclave, the resort, the resort destination, the cruise ship and besieged cities like Barcelona remains to be investigated from a political philosophical perspective. The all-inclusive resort for instance is a way for the sovereign to biopolitically organise and manage space in order to secure the control and supervision of the subjects and their mobility (Mattson, 2016). A second avenue is to in a more philosophical way attempt to unravel the guest – host paradigm, through for instance a metaphorical and symbolic refiguration of the guest and the host as untidy and in the extension reformulate hospitality as a way of being rather than something that takes place solely in a touristic context (Veijola, Molz, Pyyhtinen, Höckert, & Grit, 2014). A third direction could be centered on the tourist *per se*, as in Jansson (2018) or perhaps rather a figuration of the 'post-post tourist', a social media drenched figure that de-differentiates cultural boundaries and distinctions between mass tourism and niche tourism as well as between tourism and post-tourism.

Fourth, we propose that enclavic tourism research in particular and tourism studies in general would benefit by engaging with the spherical ontologies proposed by Sloterdijk (2016). Sloterdijk (2014, p. 195) places tourism at the altar of the capitalist way of life. From this perspective, the tourism enclave is not just a distillation of a new meta-geography of the Global South and the Global North and developmental issues therein, but also a meta-geography of the planetary polity (Sidaway, 2007). Following Sloterdijk (2016), in relation to urban tourism for instance, it would be worthwhile to examine tourism enclaves in light of *the foam* of planetary urbanism, 'tourism as part of the interior world of capital' so to speak (Sloterdijk, 2016). Airbnb and Uber are iconic examples of this. Indeed, *the foam*, as well as *the sphere* and *the bubble* draw out the dialectics of inclusion/exclusion, inside/outside, norm/exception in sharper relief than Agamben's esoteric/metaphysical musings on space, power, norm/exception, and life/death. For instance, Sloterdijk (2016) theorizes the foam as the spatial architecture of the contemporary, such that we live in our small bubbles, connected-and-encased. Translated into the tourism lexicon this would shed new insights not only on the workings of tourism enclaves, but also on current tourist practices like tourism social media and other attempts to encapsulate and de-differentiate tourist subjectivities (Munar, Gyimóthy, & Cai, 2013).

Our critical scrutiny of tourism is no stranger to tourism theory itself. Rather, a careful convocation of (what at first *sight* appear as) far-flung analogies to concepts like 'camp', and 'foam'. We hope the previous insights will hold out a vitalization of tourism theory. In any case, topological concepts offer a widening of the ontological horizon of our discipline. In our *view*, this is beneficial for tourism studies both theoretically and empirically.

Our theoretical antes are not devoid of methodological risks. The obvious one being the mapping of philosophical and metaphysical arguments on to the plain field of tourism studies. So too the transposition of philosophical concepts and arguments to the realm of the social sciences is fraught with methodological and conceptual problems. As the attempts to import Agamben's ideas and topological perspectives from the discipline of mathematics to the social sciences show – a daunting task and intellectual endeavor to partake. Yet, this in itself should not hinder sustained and critical engagements either. Indeed, is it not by challenging ourselves intellectually that we develop as academics?

Disclosure statement

No potential conflict of interest was reported by the authors.

References

Abrahamsson, C. (2012). Mathematics and space. *Environment and Planning D: Society and Space, 30*(2), 315–321. doi:10.1068/d3002int

Allen, J. (2016). *Topologies of power. Beyond territory and networks.* Abingdon: Routledge.

Agamben, G. (1998). *Homo Sacer. Sovereign power and bare life.* Stanford: Stanford University Press.

Agnew, J. (1994). The territorial trap: The geographical assumptions of International Relations Theory. *Review of International Political Economy, 1*(1), 53–80. doi:10.1080/09692299408434268

Anderson, W. (2011). Enclave tourism and its socio-economic impact in emerging destinations. *Anatolia – Anatolia, 22*(3), 361–377. doi:10.1080/13032917.2011.633041

Barba Lata, I., & Minca, C. (2016). The surface and the abyss/rethinking topology. *Environment and Planning D: Society and Space, 34*(3), 438–455. doi:10.1177/0263775815622086

Berger, S. (2010). Geopolitics of enclaves: The study of enclaves – some introductory remarks. *Geopolitics, 15*(2), 312–328. doi:10.1080/14650040903486942

Bianchi, R. W., & Stephenson, M. L. (2014). *Tourism and citizenship*. Abingdon: Routledge.

Büscher, B., & Fletcher, R. (2017). Destructive creation: Capital accumulation and the structural violence of tourism. *Journal of Sustainable Tourism, 25*(5), 651–667. doi:10.1080/09669582.2016.1159214

Buzinde, C. N., & Manuel-Navarrete, D. (2013). The social production of space in tourism enclaves: Mayan children's perceptions of tourism boundaries. *Annals of Tourism Research, 43*, 482–505. doi:10.1016/j.annals.2013.06.003

Cantillon, Z. (2018). *Resort spatiality. Reimagining sites of mass tourism*. London: Routledge.

Carlisle, S., & Jones, E. (2012). The beach enclave. A landscape of power. *Tourism Management Perspectives, 1*, 9–16. doi:10.1016/j.tmp.2011.12.003

Casey, E. (2002). *Representing place: Landscape paintings and maps*. Minneapolis, MN: University of Minnesota Press.

Cohen, E. (2006). Pai – A backpacker enclave in transition. *Tourism Recreation Research, 31*(3), 11–27. doi:10.1080/02508281.2006.11081502

Cohen, E., & Neal, M. (2012). A middle eastern Muslim tourist enclave in Bangkok. *Tourism Geographies, 14*(4), 570–598. doi:10.1080/14616688.2012.647320

Dikeç, M. (2015). *Space, politics and aesthetics*. Edinburgh: Edinburgh University Press.

Diken, B., & Laustsen, C. B. (2005). *The culture of exception: Sociology facing the camp*. Abingdon: Routledge.

Diken, B., & Laustsen, C. B. (2006). The camp. *Geografiska Annaler B, 88*(4), 443–452. doi:10.1111/j.0435-3684.2006.00232.x

Doel, M. (2000). Un-glunking geography: Spatial science after Dr. Seuss and Gilles Deleuze. In M. Crang & N. Thrift (Eds.), *Thinking space* (pp. 117–136). London: Routledge.

Edensor, T. (2000). Staging tourism: Tourists as performers. *Annals of Tourism Research, 27*(2), 322–344. doi:10.1016/S0160-7383(99)00082-1

Edensor, T. (2006). Performing tourist places. *Journal of Rural Studies, 22*(2), 243–244. doi:10.1016/j.jrurstud.2005.07.006

Edensor, T. (2007). Mundane mobilities, performances and spaces of tourism. *Social and Cultural Geography, 8*(2), 189–215.

Ek, R. (2006). Giorgio Agamben and the spatialities of the camp: An introduction. *Geografiska Annaler Series B Human Geography, 88*(4), 363–386. doi:10.1111/j.0435-3684.2006.00228.x

Ek, R., & Larson, M. (2017). Imagining the alpha male of the tourism tribe. *Anatolia – Anatolia, 28*(4), 540–552. doi:10.1080/13032917.2017.1370778

Ek, R., & Tesfahuney, M. (2016). The paradigmatic tourist. In A. M. Munar & T. Jamal (Eds.), *Tourism research paradigms: Critical and emergent knowledges* (pp. 113–129). Bingley: Emerald Press.

Elden, S. (2009). *Terror and territory. The spatial extent of territory*. Minneapolis, MN: University of Minnesota Press.

Elsrud, T. (2004). Taking time and making journeys: Narratives on Self and the Other among backpackers. Lund Dissertations in Sociology 56. Department of Sociology, Lund University.

Franklin, A. (2003). The tourist syndrome. *Tourist Studies, 3*(2), 205–217. doi:10.1177/1468797603041632

Franklin, A. (2004). Tourism as an ordering: Towards a new ontology of tourism. *Tourist Studies, 4*(3), 277–301. doi:10.1177/1468797604057328

Franklin, A. (2007). The problem with tourism theory. In I. Ateljevic, A. Pritchard, & N. Morgan (Eds.), *The critical turn in tourism studies* (pp. 131–148). Oxford: Elsevier.

Freitag, T. G. (1994). Enclave tourism development. For whom the benefits roll? *Annals of Tourism Research, 21*(3), 538–554. doi:10.1016/0160-7383(94)90119-8

Foucault, M. (2003). *Society must be defended*. London: Picador.

Giaccaria, P., & Minca, C. (2011). Topographies/topologies of the camp: Auschwitz as a spatial threshold. *Political Geography, 30*(1), 3–12. doi:10.1016/j.polgeo.2010.12.001

Gregory, D. (1994). *Geographical imaginations*. Cambridge, MA: Blackwell.

Gregory, D. (2004). *The colonial present: Afghanistan, Palestine, Iraq*. Malden: Blackwell.

Gregory, D. (2006). The black flag. Guantánamo Bay and the space of exception. *Geografiska Annaler Series B: Human Geography, 88*(4), 405–427. doi:10.1111/j.0435-3684.2006.00230.x

Hannam, K. (2008). Tourism geographies, tourism studies and the turn towards mobilities. *Geography Compass, 2*(1), 127–139. doi:10.1111/j.1749-8198.2007.00079.x

Haraway, D. (1991). *Simians, cyborgs and women: The reinvention of nature*. London: Free Association Books.

Harvey, D. (2006). Space as a key word. In N. Castree & D. Gregory (Eds.), *David Harvey: A critical reader* (pp. 270–293). Malden: Blackwell.

Heidegger, M. (1977). *The question concerning technology, and other essays*. New York, NY: Harper & Row.

Häkli, J. (2001). In the territory of knowledge: State-centered discourses and the construction of society. *Progress in Human Geography, 25*(3), 403–422. doi:10.1191/030913201680191745

Jansson, A. (2018). Rethinking post-tourism in the age of social media. *Annals of Tourism Research, 69*, 101–110. doi:10.1016/j.annals.2018.01.005

Jay, M. (1988). Scopic regimes of modernity. In H. Foster (Ed.), *Vision and visuality* (pp. 3–23). Seattle: Bay Press.

Kermath, B. M., & Thomas, R. N. (1992). Spatial dynamics of resorts. Sosúa, Dominican Republic. *Annals of Tourism Research, 19*(2), 173–190. doi:10.1016/0160-7383(92)90076-2

Kingsbury, P. (2005). Jamaican tourism and the politics of enjoyment. *Geoforum, 36*(1), 113–132. doi:10.1016/j.geoforum.2004.03.012

Klein, N. (2007). *The shock doctrine: The rise of disaster capitalism*. New York, NY: Metropolitan Books/Henry Holt.

Koopman, S. (2008). Cutting through topologies: Crossing lines at the school of Americas. *Antipode, 40*(5), 825–847. doi:10.1111/j.1467-8330.2008.00639.x

Latour, B. (1993). *We have never been modern*. Cambridge, MA: Harvard University Press.

Lefebvre, H. (1991). *The production of space*. Oxford: Basil Blackwell.

Levin, D. M. (1989). *The listening self. Personal growth, social change and the closure of metaphysics*. London: Routledge.

Levin, D. M. (1993). Introduction. In D. M. Levin (Ed.), *Modernity and the hegemony of vision* (pp. 1–29). Berkeley: University of California Press.

Lloyd, K. (2006). Catering to the backpacker: The transition of backpacker enclaves in Vietnam. *Tourism Recreation Research, 31*(3), 65–73. doi:10.1080/02508281.2006.11081506

Löfgren, O. (1999). *On holiday. A history of vacationing*. Berkeley: University of California Press.

MacCannell, D. (1999). *The tourist. A new theory of the leisure class*. Berkeley: The University of California Press.

Manuel-Navarrete, D. (2016). Boundary-work and sustainability in tourism enclaves. *Journal of Sustainable Tourism, 24*(4), 507–526. doi:10.1080/09669582.2015.1081599

Martin, L., & Secor, A. J. (2014). Towards a post-mathematical topology. *Progress in Human Geography, 38*(3), 420–438. doi:10.1177/0309132513508209

Marx, K. (2002). Grundrisse: Foundations of the critique of political economy. In N. W. Biggart (Eds.), Readings in economic sociology (pp. 18–23). Oxford: Blackwell. doi:10.1002/9780470755679.ch2

Massey, D. (2005). *For space*. London: Sage.

Mattson, K. (2016). Holiday utopias: The gendered and racialized spaces of all-inclusive tourism. In M. Tesfahuney & K. Schough (Eds.), *Privileged mobilities: Tourism as world-ordering* (pp. 67–80). Newcastle upon Tyne: Cambridge Scholars Publishing.

Mbaiwa, J. E. (2005). Enclave tourism and its socio-economic impact in the Okavango Delta. *Tourism Management, 26*(2), 157–172. doi:10.1016/j.tourman.2003.11.005

Minca, C. (2005). The return of the camp. *Progress in Human Geography, 29*(4), 405–412. doi:10.1191/0309132505ph557xx

Minca, C. (2006). Giorgio Agamben and the new biopolitical nomos. *Geografiska Annaler B, 88*(4), 387–403. doi:10.1111/j.0435-3684.2006.00229.x

Minca, C. (2007). Agamben's geographies of modernity. *Political Geography*, *26*(1), 78–97. doi:10.1016/j.polgeo.2006.08.010

Minca, C. (2009). The island: Work, tourism and the biopolitical. *Tourist Studies*, *9*(2), 88–108. doi:10.1177/1468797609360599

Mitchell, T. (1989). *The world as exhibition*. Cambridge: Cambridge University Press.

Mostafanezhad, M. (2018). The geopolitical turn in tourism geographies. *Tourism Geographies*, *20*(2), 343–346. doi:10.1080/14616688.2018.1434820

Mouffe, C. (2005). *On the political*. London: Routledge.

Munar, A. M., Gyimóthy, S., & Cai, L. eds. (2013). *Tourism social media: Transformations in identity, community and culture*. Bingley: Emerald.

Murdoch, J. (2006). *Post-structuralist geography: A guide to relational space*. London: Sage.

Naidoo, P., & Sharpley, R. (2016). Local perceptions of the relative contributions of enclave tourism and agritourism to community well-being: The case of Mauritius. *Journal of Destination Marketing & Management*, *5*, 16–25. doi:10.1016/j.jdmm.2015.11.002

Nail, T. (2016). *Theory of the border*. New York, NY: Oxford University Press.

Nancy, J.-L. (2000). *Being singular plural*. Stanford: Stanford University Press.

Netz, R. (2004). *Barbed wire: An ecology of modernity*. Middletown, CT: Wesleyan University Press.

Nietzsche, F. (1996). *Human, all too human*. Cambridge: Cambridge University Press.

Nilsson, M., & Tesfahuney, M. (2018). Home – tourism's uncanny. *Current Issues in Tourism*, *22*(15), 1783–1788. doi:10.1080/13683500.2018.1441269

Norris, A. ed. (2005). *Politics, metaphysics, death. Essays on Giorgio Agamben's Homo Sacer*. Durham: Duke University Press.

Pernecky, T. (2010). The being of tourism. *Journal of Tourism and Peace Research*, *1*(1), 1–15.

Ranciére, J. (1999). *Disagreement. Politics and philosophy*. Minneapolis, MN: University of Minnesota Press.

Rodaway, P. (1994). *Sensuous geographies. Body, sense and place*. London: Routledge.

Roelofsen, M., & Minca, C. (2018). The superhost. Biopolitics, home and community in the *Airbnb* dream-world of global hospitality. *Geoforum*, *91*, 170–181. doi:10.1016/j.geoforum.2018.02.021

Rojek, C. (1998). Tourism and citizenship. *International Journal of Cultural Policy*, *4*(2), 291–310. doi:10.1080/10286639809358077

Rutherford, S. (2011). *Governing the wild. Ecotours of power*. Minneapolis, MN: The University of Minnesota Press.

Saarinen, J. (2017). Enclavic tourism spaces: Territorialisation and bordering in tourism destination development and planning. *Tourism Geographies*, *19*(3), 425–437. doi:10.1080/14616688.2016.1258433

Sassen, S. (2006). *Territory, authorities, rights: From medieval to global assemblages*. Princeton: Princeton University Press.

Shaw, B. J., & Shaw, G. (1999). Sun, sand and sales': Enclave tourism and local entrepreneurship in Indonesia. *Current Issues in Tourism*, *2*(1), 68–81. doi:10.1080/13683509908667844

Shields, R. (2013). *Spatial questions: Cultural topologies and social spatialisation*. London: Sage.

Sidaway, J. D. (2007). Enclave space: A new metageography of development. *Area*, *39*(3), 331–339. doi:10.1111/j.1475-4762.2007.00757.x

Simpson, T. (2016). Tourist utopias: Biopolitics and the genealogy of the post-world tourist city. *Current Issues in Tourism*, *19*(1), 27–59. doi:10.1080/13683500.2015.1005579

Sloterdijk, P. (2014). *In the world interior to capital. Towards a philosophical theory of globalisation*. Cambridge: John Wiley.

Sloterdijk, P. (2016). *Foams. Spheres III*. South Pasadena, CA: Semiotext(e).

Smith, N. (2003). *American empire: Roosevelt's geographer and the prelude to globalization*. Berkeley: University of California Press.

Swyngedouw, E. (2018). *Promises of the political: Insurgent cities in a post-political environment*. Cambridge, MA: MIT Press.

Tesfahuney, M. (2016). Who is the tourist? Reflection on the ideal typical subject of our time. In M. Tesfahuney & K. Schough (Eds.), *Privileged mobilities. Tourism as world ordering* (pp. 93–105). UK: Cambridge Scholars Press.

Torres, R. M., & Momsen, J. D. (2005). Gringolandia. The construction of a new tourist space in Mexico. *Annals of the Association of American Geographers, 95*(2), 314–335. doi:10.1111/j.1467-8306.2005.00462.x

Veijola, S., Molz, G. J., Pyyhtinen, O., Höckert, E., & Grit, A. (2014). *Disruptive tourism and its untidy guests. Alternative ontologies for future hospitalities*. Basingstoke: Palgrave Macmillan.

Wall, T. C. (2005). Au hazard. In A. Norris (Ed.), *Politics, metaphysics, death. Essays on Giorgio Agamben's Homo Sacer* (pp. 31–48). Durham: Duke University Press.

Wilson, J., & Richards, G. (2008). Suspending reality: An exploration of enclaves and the backpacker experience. *Current Issues in Tourism, 11*(2), 187–202. doi:10.2167/cit371.0

Žižek, S. (1996). *The sublime object of ideology*. London: Verso.

Žižek, S. (1999). *The ticklish subject: The absent centre of political ontology*. New York: Verso.

Conclusion: Geographies of exclusive spaces in tourism – Future issues

Jarkko Saarinen and Sandra Wall-Reinius

Introduction

This collection discussed and overviewed the nature and basic ideas related to enclaves and enclavisation processes in tourism development and tourist consumption. The aim was to deepen our understanding of how these distinct exclusive tourism spaces are structured and transformed and what impact and implications they have for research, tourism and local development, governance and regional economies, well-being and sustainability. In general, tourism enclaves can be understood as exclusively planned spaces that contain the facilities and services needed for visitors who have limited opportunities and/or desire to leave the enclave tourism space (Saarinen, 2017). In tourism literature, such spaces refer to all-inclusive resorts, gated recreation and resort communities, privatised beaches, cruise liners or foreign cruise liner-owned islands, for example, and they are usually outcomes of a non-locally driven and governed transnational market economy (see Britton, 1991; Shaw & Shaw, 1999).

While being externally governed and disconnected from local realities, tourism enclaves are deeply anchored to context-specific planning and development practices, which have significant impacts on local realities, ways of living, and well-being. Local people's access to enclaves, for example, may be regulated, and enclaves are usually physically and functionally separated from the surrounding communities (see Buzinde & Manuel-Navarrete, 2013; Torres & Momsen, 2005). Indeed, enclaves are separated in many ways from their surrounding areas, but the nature and degree of differentiation between the hosts and guests and their economies and cultures vary depending on the context. Furthermore, the economic benefits of tourism do not easily diffuse outside the enclave (Anderson, 2011; Britton, 1991). Thus, these all-inclusive tourism products and environments "can turn out to be all-exclusive spaces for local communities in development" (Saarinen, 2017, p. 433). This selective inclusiveness and exclusiveness can lead to highly unsustainable outcomes in tourism planning and development.

The introduction of this collection overviewed the basic premises of enclave tourism and related concepts and frameworks. It also introduced and concluded the individual papers of the book, and discussed some of the future research needs on tourism enclaves. Here, we briefly focus on some of the most common, but also some of the recently emerging elements and processes that may influence the evolution of tourism enclaves and their nature in the future.

Exclusive spaces: Tourism and changing destinations

Intensified cycles of change by globalisation

The changing character of tourism and its local impacts have been theorised and modelled in various ways. Butler (1980, p. 5) stated in his seminal work on the tourism area life cycle (TALC): "there can be little doubt that tourist areas are dynamic, that they evolve and change over time." Undeniably, constant change is an elementary feature of tourism, and this change causes various impacts and processes in tourism destinations and the whole tourism system (see Leiper, 1979, 1990). This changing nature of destinations has interested tourism geographers for a long time. Gilbert (1939, 1949), for example, studied the development of seaside and inland resorts in Great Britain by focusing on the issues lying behind their change. These included elements such as transportation systems, medical knowledge and world politics. Christaller (1964), who is better known for his systematic approaches to land use patterns, and especially for Central Place theory, also made efforts to study the changing spatial organisation of tourism destinations. For him, tourism was a typical phenomenon for peripheral areas where tourism evolves from individual, unorganised and small-scale activity to institutionalised and large-scale operations with intensive marketing, competition and package tours.

These early depictions of tourism and its changing characteristics were further developed by Stansfield (1978) and, eventually, by Butler (1980). For the latter, tourism was a dynamic force that evolves through specific stages from a highly embedded small-scale individual tourism and close host-guest relations (exploration and involvement stages) to a development stage, and finally to mass-scale tourism destination operations (consolidation/stagnation stages) with formalised or limited host-guest relations. After the stagnation stage, Butler considered that the cycle can start again (rejuvenation), or it could decline due to earlier overdevelopment resulting in lost attractiveness and increasing competition from newly evolving and (still) more attractive destinations (see Strapp, 1988). This process and its specific stages and the overall applicability of the evolution cycle in theory and practice have been widely debated in tourism literature (see Agarwal, 2002; Butler, 2015; Cooper, 1989; Haywood, 1986, 1992; Saarinen, 2004).

Here, the interest is not on whether the reality follows the conceptual stages and the (pedagogical) curve form of the model. At a general level, the evolution cycle demonstrates a tourism development process from locally integrated operations towards a highly specialised and separated tourism economy, which may have limited connections to surrounding socio-economic structures. In this respect, the model corresponds with the general idea of the enclavisation process as discussed in the contributions to this book. This indicates that the aspect of enclavisation may characterise all kinds of tourism development processes, in general, making the phenomenon much more typical and embedded to tourism than we usually consider in tourism studies. Obviously, not all mass tourism destinations become enclaves and the degree of bordering between tourism and local realities – guests and hosts – can vary greatly from symbolic, visible and blurred to material, structural and unequal manifestations in socio-political and economic relations in destination areas and communities (Britton, 1982, 1991; Judd, 1999; Saarinen, 2017).

While strict tourism enclaves with clear territorial characteristics and strong bordering mechanisms are considered exceptions in tourism, the process itself creates distinctions between tourism and localities that are built-in to tourism development (see Butler, 1980; Timothy, Saarinen, & Viken, 2016). What makes this interesting and probably increasingly problematic for local and regional economies with a tourism development component is the intensified global-scale relationality and dependency in socio-economic relations (Brohman, 1996; Shaw & Shaw, 1999). There is nothing new in globalisation and non-locally driven development in tourism, but recent decades with deepening policies and practices linked to neoliberal capitalism have intensified the changing characteristics of global tourism. As a result, destinations are transformed much more rapidly, and with a non-local basis, than ever before (see Bianchi, 2006; Saarinen, Rogerson, & Hall, 2017). Thus, the evolution cycle of a tourism destination from early stages to stagnation may have taken numerous decades before, but due to easier access to financial capital, better global transportation systems and the digitalisation of marketing and bookings, for example, the cycle can be progressed much faster. As a result, globalisation has increased human mobility and how places, agencies and economies are connected ever more interdependently on each other (see Rumford, 2008, 2012; Storper, 1997). Paradoxically, intensified globalisation and transnational mobilities, including the democratisation of tourism in a form of mass and over-tourism, have probably increased enclavisation in tourism (see Milano, Novelli, & Cheer, 2019; Saarinen, 2017, 2019). In addition to the general globalisation process and tourism being more affordable to people than ever before, however, there are other contemporary elements that potentially increase the formation of exclusive spaces for tourism. At the same time, there are also processes that may work against enclavisation. For the former, we use COVID-19 as an example of a potentially intensifying element working for enclave tourism in future. For the latter, an obvious perspective opens up based on a sustainability discourse, which has guided tourism development policies since the 1990s. However, we estimate that it will eventually turn towards much more concrete action in the future, which will be based on an urgent need for firmer climate change policies and actions, for example, and how these will influence the overall global tourism industry and its patterns and growth-oriented development thinking. These two perspectives can also be seen as being interconnected: as noted by Ioannides and Gyimóthy (2020), the COVID-19 crisis in tourism has given us an opportunity to rethink and select a new direction for tourism based on a more sustainable development path (see Gössling, Scott, & Hall, 2020; Rogerson & Baum, 2020). Thus, the COVID-19 crisis can both intensify and lower some of the bordering elements in tourism enclaves.

COVID-19 beyond lockdowns

COVID-19 has had a major impact on the global tourism industry. The pandemic has almost frozen global human mobility, and it has also created a greater awareness of how important tourism experiences and consumption are for people and local economies and communities around the world (see Baum et al., 2020; Jones & Comfort, 2020). Travel and mobility restrictions have been highly unexpected, and the social and economic effects on the tourism and hospitality industry and tourism-dependent communities and economies have been disastrous (Ioannides & Gyimóthy, 2020). With regard to the way forward, there are two narratives. Both highlight the seriousness of the COVID-19 crisis for the current global tourism industry, but they differ in terms of what comes or should come next. *A resilience school of thought* emphasises the historical record of tourism to cope with and adapt to change and sudden shocks, such as oil and political crises, SARS and MERS, and the industry's relatively fast and good capacity to turn towards the pre-crisis growth path (see Brouder, 2020). This resilience perspective is typically expressed by governmental and policy-making sectors and the industry itself (see Gössling, Scott, & Hall, 2020). In contrast to this, *a readjustment school of thought* emphasises the need to rethink the growth- and volume-driven orientation of tourism and neoliberal capitalism in general, and interprets the watershed nature of COVID-19 by stressing its irreversible impact that will fundamentally change the nature of tourism and human mobility in the future (see Higgins-Desbiolles, 2020; Sigala, 2020). Obviously, this prediction, hopeful wish or an urgent call for radical change in the global tourism economy towards a sustainable (or even degrowth) development path, is based on critical tourism thinking in academia (Bianchi, 2009, 2020).

Regardless of the school of thought, the COVID-19 crisis has had a significant impact on contemporary tourism enclaves. Obviously, due to total lockdowns or highly restricted domestic and/or international travelling in many countries, all kinds of tourist destinations have limited operations. In addition, there are also new kinds of social conflicts emerging in destination communities between 'us' and 'them' (see Jamal & Budke, 2020). Tourism enclaves, in particular, are in a catastrophic situation economically: compared to many other types of tourism spaces they have a limited capacity to attract alternative domestic tourist segments or work towards local consumption, as has happened in some other types of tourism, such as nature-based tourism, during the pandemic. In an extreme situation, their operations are fully closed. This is the case with cruise liners (see Weaver, and Paananen & Minoia in this book), for example, and coastal resort environments that have been very dependent on cruise passenger visits (see Renaud, 2020; Weaver, 2005). As noted by Gössling, Scott and Hall (2020, p. 12) 'it is unlikely that cruise ships can sail again before a vaccine is found or unless passengers can be tested before boarding'.

At the same time, however, the COVID-19 crisis may also strengthen or even create new exclusive spaces in tourism. Enclaves with strict bordering mechanisms and, thus, limited or non-existing social contact with the surrounding environments and local communities may be perceived as safe by visitors, as social distancing is easier to manage. It is well-known from previous research that all-inclusive resorts are perceived by tourists to be safe, secure and controlled, and personal safety reasons are important motives for tourists to choose all-inclusive resorts (Wong & Kwong, 2004; see also Carlisle & Jones, 2012; Wall-Reinius, Ioannides, & Zampoukos in this book). Some destinations have even targeted wealthy tourists who are willing to buy or rent a private island resort for their own use as 'safe havens' (see McKeever, 2020; Smyth, 2020). This 'enclavisation for safety' is supported by existing individualisation trends of Western lifestyles and consumption patterns (see Beck & Beck-Gernsheim, 2002), which has also resulted in various forms of luxury tourism, for example (Brenner & Aguilar, 2002). Thus, the tourism enclaves become integral parts of the geographies of the super-rich (see Hay, 2013, and dell'Agnese in this book).

In addition, COVID-19 policies have resulted in plans for creating enclavic structures to destinations that may not have been previously characterised as firmly exclusive spaces separated from surrounding areas. In Finnish Lapland, for example, the regional tourism industry proposed in September 2020, with support from the region's health care district, to open up national borders to a wide spectrum of international Christmas tourism visitors who would be 'channeled' from their point of entry to the destination resort and selected tourist activity locations (e.g. husky safaris, visit to Santa Claus, snowmobiling) without any contact with local people and environments (see Yle, 2020). It is argued that this plan will help save the industry and its employees over the winter season. It would also result in a similar discouragement but much firmer bordering for tourists to visit areas and/or attractions outside the designated core resort areas as described by Wall-Reinius, Ioannides and Zampoukos (in this book).

Deeper call for sustainability: Sustainable enclaves?

As noted, the readjustment school of thought highlights an alternative sustainable development path that the COVID-19 crisis creates or forces the tourism industry and its governance actors to think about and apply. However, this rethinking or resetting may involve very different aims and degrees of readjustments, both ideologically and practically. There have been recent calls for a reset to the global capitalistic system. Wolf (2019), for example, stated that the reason the current economic system is not delivering the expected growth and benefits lies with the rise of so-called rentier capitalism (see Saarinen, 2021). What he sees as necessary is "a dynamic capitalist economy that gives everybody a justified belief that they can share in the benefits" (Wolf, 2019). Similarly, Moyo (2018) has called for a retooling of the economy. Still, both Moyo (2018) and Wolf (2019), among many other economic commentators, emphasise the key role of growth for our future: what they call for is a more responsible and better governed capitalistic economy than the current one.

However, there are other perspectives calling for a more radical readjustment of our economic thinking and system. There the quest is not to fine-tune the capitalistic mode of production and consumption with a different kind of responsible growth aspect. From this perspective the aim is to find an alternative replacement for growth-oriented capitalism (Kerschner, 2010). This radical shift focuses on the possibilities of degrowth (Fournier, 2008; Koch, 2020), referring to a situation in which the economic wealth produced does not increase or it may even decrease (Kerschner, 2010; Saarinen, 2021). In the context of tourism, Hall (2009) has stressed the need for the industry to seek alternative approaches to volume-based economic development by integrating sustainable development with degrowth thinking.

Although these calls for more responsible capitalism or alternative radical degrowth approaches represent very different degrees of readjustment for the tourism economy and its future development paths, they share a common emphasis on sustainability thinking (see Saarinen, 2021). Thus, in general, the COVID-19 crisis has provided us with an opportunity to re-evaluate the logics of tourism growth and move to implement sustainability to the practices and governance of the tourism industry as noted by Ioannides and Gyimóthy (2020). With regard to the sustainability emphasis, the challenge has not been a low recognition of the need to have a more sustainable tourism industry (see Hall, 2011, 2019). The problem has been its implementation, i.e. how to move beyond the rhetoric. This problem, however, may characterise sustainable development policies and processes in general, not only in tourism. Indeed, as stated by Utting (2015, p. 1) the core elements and ideas of sustainability "often got lost in translation" when we are 'doing development' (see Saarinen, 2018).

The industry could not have foreseen the COVID-19 crisis. In contrast to this, the accelerating global climate change has been widely realised and discussed, and there is an increasing understanding that a stronger call and practical implementations for sustainability governance at the tourism system level are needed for climate change resilience and related mitigation and adaptation (see Gössling, Scott, & Hall, 2020; Jamal & Camargo, 2018; Saarinen & Gill, 2019). Firmer sustainability governance and the emphasis on readjusting tourism development thinking by having a more responsible, inclusive and just tourism economy will potentially challenge the bordering mechanisms of tourism enclaves in the future (Prayag, 2015; Scheyvens & Biddulph, 2018). Sustainable tourism cannot be implemented by managing environmental issues alone (Butler, 1999). Thus, the predicted greater demand for sustainable development implementation in tourism will not only focus on carbon emission control, for example, but also wider environmental justice issues and socio-political and economic cost- and benefit-sharing elements. All this should empower local actors and economies to participate more effectively in the tourism industry and the destination spaces it utilises in the future, which is highly relevant in matters relating to enclave tourism.

However, it is highly unlikely that tourism enclaves will disappear in the near future. First, as indicated earlier, enclavisation seems to be a characterising feature for the tourism development process in general. Bordering practices can be lowered but human territorialisation may not easily be reduced (see Rumford, 2008; Sack, 1986) based on better sustainability governance in the tourism sector alone. In addition, as noted by Butler (1990, p. 40), "many people seem to enjoy being a mass tourist." For consumers, enclaves are easy all-inclusive products to consume with a single price paid in the tourists' home regions covering all or most of their expenses (Saarinen, 2017). As a result, all-inclusive tourism spaces have "become a dominant travel choice to both developed and developing destinations" (Farmaki, Georgiou, & Christou, 2017, p. 496).

This creates an interesting but highly challenging task for the future tourism industry on how to plan and manage 'sustainable enclaves' that would satisfy both the hosts' and guests' needs while leading separate lives in the destination spaces. In this respect, successful community-based natural resource management (CBNM) models and cases with a tourism development component may serve as examples (see Blaikie, 2006; Ngoni & Saarinen, 2020), with some implementation challenges to mass-scale tourism products likely (see Farmaki, Georgiou, & Christou, 2017).

Concluding remarks

In the current global economic circumstances, tourism enclaves are products of a non-locally driven neoliberal market economy characterised by a diverse set of bordering practices and limited economic connections with surrounding social realities in destination communities. In this respect, the enclaves are economically and politically more connected and dependent on international- or national-scale actors and economic conditions (Britton, 1991). Based on the contributions of this volume, tourism enclaves can be understood as purposefully planned, developed and managed spaces that involve the facilities and activities needed for guests to be entertained and served, while locals' access to these exclusive spaces are often somehow limited. Thus, there can be clear power imbalances and elements of inequality involved with tourism enclaves.

Former studies on tourism enclaves have been based on various approaches, including dependency theory, political economy and political ecology approaches (Britton, 1991; Fregonese & Ramadan, 2015; Manuel-Navarrete, 2016; Mbaiwa, 2005; Mosedale, 2011; Nepal & Saarinen, 2016), as well as analysis of socio-spatial relations and representations of space (see Buzinde & Manuel-Navarrete, 2013; Carlisle & Jones, 2012; Cohen, 2006; Dürr, 2012; Nunkoo, & Ramkissoon, 2016; Saarinen & Niskala, 2009; Torres & Momsen, 2005). In addition to the research needs expressed in the introduction to this collection, the current COVID-19 crisis, together with the calls for stronger sustainability governance in global tourism, is creating new research needs and challenges. There are also increasing calls for a total rethink on research priorities and topics in tourism (Gössling et al., 2020; Prideaux, Thompson, & Pabel, 2020; Sigala, 2020). In the African context, Rogerson and Baum (2020, p. 7) have also highlighted the need for post-COVID-19 tourism development and research "to become more aligned with SDGs." This imperative should also apply to the Global North (and obviously the rest of the Global South) (see United Nations, 2015).

In respect of tourism enclaves, post-COVID-19 research needs remain the same as before, in principle: how human territorialisation and border-making operates in the global-local nexus in tourism; the socio-economic, political and environmental consequences of enclavisation; policy-making and planning responses, power relations, social and environmental justice and citizen participation; and benefit- and cost-sharing strategies for sustainability governance in destination development, for example. However, due to the COVID-19 crisis and potentially changing patterns of global tourism, we also need more nuanced and future-oriented analysis of the elements of destination resilience and sustainability in changing conditions. This is needed for future economic and societal transformation needs, driven by evolving global- or regional-scale risks and shocks based on accelerating climate change impacts or the next epidemic/pandemic, towards more inclusive tourism development models for destination communities.

References

Agarwal, S. (2002). Restructuring seaside tourism: The resort lifecyle. *Annals of Tourism Research, 29*(1), 25–55.

Anderson, W. (2011). Enclave tourism and its socio-economic impact in emerging destinations. *Anatolia, 22*(3), 361–377.

Baum, T., Mooney, S. K. K., Robinson, R. N. S., & Solnet, D. (2020). COVID-19's impact on the hospitality workforce – new crisis or amplification of the norm? *International Journal of Contemporary Hospitality Management, 32*(9), 2813–2829. doi:10.1108/IJCHM-04-2020-0314

Beck, U., & Beck-Gernsheim, E. (2002). *Individualization: Institutionalized individualism and its social and political consequences*. Thousand Oaks, CA: Sage.

Bianchi, R. (2006). Tourism and the globalization of fear: Analyzing the politics of risk and (in)security in global travel. *Tourism and Hospitality Research, 7*(1), 64–74. doi:10.1057/palgrave.thr.6050028

Bianchi, R. (2009). The 'critical turn' in tourism studies: A radical critique. *Tourism Geographies, 11*, 484–504. doi:10.1080/14616680903262653

Bianchi, R. (2020). COVID-19 and the potential for a radical transformation of tourism? *ATLAS Tourism and Leisure Review, 2*, 80–86.

Blaikie, P. (2006). Is small really beautiful? Community-based natural resource management in Malawi and Botswana. *World Development, 34*(11), 1942–1957.

Brenner, L., & Aguilar, A. G. (2002). Luxury tourism and regional economic development in Mexico. *The Professional Geographer, 54*(4), 500–520. doi:10.1111/0033-0124.00346

Britton, S. (1982). The political economy of tourism in the Third World. *Annals of Tourism Research, 9*, 331–358. doi:10.1016/0160-7383(82)90018-4

Britton, S. (1991). Tourism, capital, and place: Towards a critical geography of tourism. *Environment and Planning D: Society and Space, 9*, 451–478. doi:10.1068/d09045

Brohman, J. (1996). New directions in tourism for third world development. *Annals of Tourism Research, 23*, 48–70. doi:10.1016/0160-7383(95)00043-7

Brouder, P. (2020). Reset redux: Possible evolutionary pathways towards the transformation of tourism in a COVID-19 world. *Tourism Geographies*. doi:10.1080/14616688.2020.1760928

Butler, R. (1980). The concepts of a tourist area cycle of evolution: implications for management of resources. *Canadian Geographer, 24*, 5–12. doi:10.1111/j.1541-0064.1980.tb00970.x

Butler, R. (1990). Alternative tourism: Pious hope or Trojan Horse? *Journal of Travel Research, 28*, 40–45. doi:10.1177/004728759002800310

Butler, R. (1999). Sustainable tourism: A state-of-the-art review. *Tourism Geographies, 1*(1), 7–25. doi:10.1080/14616689908721291

Butler, R. (2015). The evolution of tourism and tourism research. *Tourism Recreation Research, 40*(1), 16–27.

Buzinde, C. N., & Manuel-Navarrete, D. (2013). The social production of space in tourism enclaves: Mayan children's perceptions of tourism boundaries. *Annals of Tourism Research, 43*, 482–505. doi:10.1016/j.annals.2013.06.003

Carlisle, S., & Jones, E. (2012). The beach enclave: A landscape of power. *Tourism Management Perspectives, 1*, 9–16.

Cohen, E. (2006). Pai - A backpacker enclave in transition. *Tourism Recreation Research, 31*, 11–27. doi: 10.1080/02508281.2006.11081502

Christaller, W. (1964). *Some considerations of tourism location in Europe: The peripheral regions-under development countries-recreation areas*. Regional Science Association; papers XII, Lund Congress. pp. 95–105.

Cooper, C. (1989). Tourist product cycle. In S. Witt & L. Moutinho (Eds.), *Tourism marketing and management handbook* (pp. 577–581). London: Prentice Hall.

Dürr, E. (2012). Encounters over garbage: Tourists and lifestyle migrants in Mexico. *Tourism Geographies, 12*, 339–355. doi:10.1080/14616688.2012.633217

Farmaki, A., Georgiou, M., & Christou, P. (2017). Growth and impacts of all-inclusive holiday packages: Echoed form the industry. *Tourism Planning & Development, 14*, 483–502.

Fournier, V. (2008). Escaping from the economy: The politics of degrowth. *International Journal of Sociology and Social Policy, 28*(11/12), 528–545. doi:10.1108/01443330810915233

Fregonese, S., & Ramadan, A. (2015). Hotel geopolitics: A research agenda. *Geopolitics, 20*, 793–813.

Gilbert, E. W. (1939). The growth of island and seaside health resorts in England. *Scottish Geographical Magazine, 55*(1), 16–35.

Gilbert, E. W. (1949). The growth of Brighton. *The Geographical Journal, 114*(1–3), 30–52.

Gössling, S, Scott, D., & Hall, C. M. (2020). Pandemics, tourism and global change: A rapid assessment of COVID-19. *Journal of Sustainable Tourism*. doi:10.1080/09669582.2020.1758708

Hall, C. M. (2009). Degrowing tourism: Décroissance, sustainable consumption and steady-state tourism. *Anatolia, 20*(1), 46–61. doi:10.1080/13032917.2009.10518894

Hall, C. M. (2011). Policy learning and policy failure in sustainable tourism governance: From first- and second-order to third-order change? *Journal of Sustainable Tourism, 19*(4–5), 649–671. doi:10.1080/09669582.2011.555555

Hall, C. M. (2019). Constructing sustainable tourism development: The 2030 agenda and the managerial ecology of sustainable tourism. *Journal of Sustainable Tourism, 27*(7), 1044–1060. doi:10.1080/09669582.2018.1560456

Hay, I. (Ed.). (2013). *Geographies of the super-rich*. Cheltenham: Edward Elgar Publishing.

Haywood, K. M. (1986). Can the tourist-area life cycle be made operational? *Tourism Management, 7*(3), 154–167.

Haywood, K. M. (1992). Revisiting resort cycle. *Annals of Tourism Research, 19*(2), 351–354.

Higgins-Desbiolles, F. (2020). Socialising tourism for social and ecological justice after COVID-19. *Tourism Geographies, 22*(3), 610–623.

Ioannides, D., & Gyimóthy, S. (2020). TheCOVID-19 crisis as an opportunity for escaping the unsustainable global tourism path. *Tourism Geographies, 22*(3), 624–632.

Jamal, T., & Budke, C. (2020). Tourism in a world with pandemics: Local-global responsibility and action. *Journal of Tourism Futures*, 1–8. doi:10.1108/JTF-02-2020-0014

Jamal, T., & Camargo, B. A. (2018). Tourism governance and policy: Whither justice? *Tourism Management Perspectives, 25*, 205–208.

Jones, P., & Comfort, D. (2020). The COVID-19 crisis and sustainability in the hospitality industry. *International Journal of Contemporary Hospitality Management*, 32(10), 3037–3050. https://doi.org/10.1108/IJCHM-04-2020-0357

Judd, D. (1999). Constructing the tourist bubble. In D. Judd & S. Fainstein (Eds.), *The tourist city* (pp. 35–53). New Haven, CT: Yale University Press.

Kerschner, C. (2010). Economic de-growth vs. steady-state economy. *Journal of Cleaner Production, 18*(6), 544–555. doi:10.1016/j.jclepro.2009.10.019

Koch, M. (2020). The state in the transformation to a sustainable postgrowth economy. *Environmental Politics, 29*(1), 115–133. doi:10.1080/09644016.2019.1684738

Leiper, N. (1979). The framework of tourism. *Annals of Tourism Research, 6*(4), 390–407.

Leiper, N. (1990). Tourist attraction system. *Annals of Tourism Research, 17*(3), 367–384.

Manuel-Navarrete, D. (2016). Boundary-work and sustainability in tourism enclaves. *Journal of Sustainable Tourism, 24*(4), 507–526.

Mbaiwa, J. E. (2005). Enclave tourism and its socio-economic impacts in the Okavango Delta, Botswana. *Tourism Management, 26*, 157–172.

McKeever, V. (2020). Luxury bunkers and private islands —How the rich are self-isolating from the coronavirus. CNBC Make it, 27 March 2020. Retrieved from https://www.cnbc.com/2020/03/27/coronavirus-how-the-rich-are-self-isolating.html

Milano, C., Novelli, M., & Cheer, J. M. (2019). Overtourism and tourismphobia: A journey through four decades of tourism development, planning and local concerns. *Tourism Planning & Development*. doi:10.1080/21568316.2019.1599604

Mosedale, J. (Ed.). (2011). *The political economy of tourism: A critical perspective*. London and New York: Routledge.

Moyo, D. (2018). *Edge of chaos: Why democracy is failing to deliver economic growth and how to fix it*. New York: Basic Civitas Books.

Ngoni, S., & Saarinen, J. (2020). Community perceptions on the benefits and challenges of community-based natural resources management in Zimbabwe. *Development Southern Africa*. doi:10.1080/0376835X.2020.1796599

Nepal J., & Saarinen, J. (Eds.). (2016). *Political ecology and tourism*. London: Routledge.

Nunkoo, R., & Ramkissoon, H. (2016). Stakeholders' views of enclave tourism: A grounded theory approach. *Journal of Hospitality & Tourism Research, 40*, 557–588.

Prayag, G. (2015). Beyond enclave tourism: The road to sustainability. In N. Morpeth & H. Yan (Eds.), *Planning for tourism: Towards a sustainable future* (pp. 240–258). Oxfordshire: CABI.

Prideaux, B., Thompson, M., & Pabel, A. (2020). Lessons from COVID-19 can prepare global tourism for the economic transformation needed to combat climate change. *Tourism Geographies, 22*(3), 667–678.

Renaud, L. (2020). Reconsidering global mobility – distancing from mass cruise tourism in the aftermath of COVID-19. *Tourism Geographies, 22*(3), 679–689.

Rogerson, C. M., & Baum, T. (2020). COVID-19 and African tourism research agendas. *Development Southern Africa, 37*(5), 727–741.

Rumford, C. (2008). *Citizens and borderwork in contemporary Europe*. London: Routledge.

Rumford, C. (2012). Towards a multiperspectival study of borders. *Geopolitics, 17*, 887–902. doi:10.1080/14650045.2012.660584

Saarinen, J. (2004). 'Destinations in change': The transformation process of tourist destinations. *Tourist Studies, 4*, 161–179. doi:10.1177/1468797604054381

Saarinen, J. (2017). Enclavic tourism spaces: Territorialization and bordering in tourism destination development and planning. *Tourism Geographies, 19*, 425–437.

Saarinen, J. (2018). Beyond growth thinking: The need to revisit sustainable development in tourism. *Tourism Geographies, 20*(2), 337–340. doi:10.1080/14616688.2018.1434817

Saarinen, J. (2019). Communities and sustainable tourism development: Community impacts and local benefit creation tourism. In S. F. McCool & K. Bosak (Eds.), *A research agenda for sustainable tourism* (pp. 206–222). Cheltenham: Edward Elgar Publishing.

Saarinen, J. (forthcoming). Sustainable growth in tourism? Rethinking and resetting sustainable tourism for development. In: C. M. Hall, L. Lundmark, & J. Zhang (Eds.), *Degrowth and tourism: New perspectives on tourism entrepreneurship, destinations and policy*. London: Routledge, (in press).

Saarinen, J., & Gill, A. M. (2019). Tourism, resilience and governance strategies in the transition towards sustainability. In J. Saarinen & A. M. Gill (Eds.), *Resilient destinations: Governance strategies in the transition towards sustainability in tourism* (pp. 15–33). London: Routledge.

Saarinen, J., & Niskala, M. (2009). Local culture and regional development: The role of OvaHimba in Namibian Tourism. In P. Hottola (Ed.), *Tourism strategies and local responses in Southern Africa* (pp. 61–72). Wallingford: CABI Publishing.

Saarinen, J., Rogerson, C., & Hall, C. M. (2017). Geographies of tourism development and planning. *Tourism Geographies, 19*, 307–317.

Sack, R. (1986). *Human territoriality: Its theory and history*. Cambridge: Cambridge University Press.

Scheyvens, R., & Biddulph, R. (2018). Inclusive tourism development. *Tourism Geographies, 20*, 589–609. doi:10.1080/14616688.2017.1381985

Shaw, B. J., & Shaw, G. (1999). 'Sun, sand and sales': Enclave tourism and local entrepreneurship in Indonesia. *Current Issues in Tourism, 2*, 68–81.

Sigala, M. (2020). Tourism and COVID-19: Impacts and implications for advancing and resetting industry and research. *Journal of Business Research*. doi:10.1016j.jbusres.2020.06.015

Smyth, J. (2020). Wealthy buyers snap up 'safe haven' private islands to flee pandemic. Financial Times, 24 July. Retrieved from https://www.ft.com/content/40e54149-7483-41b5-85cf-36ff200aafc6

Stansfield, C. A. (1978). Atlantic City and the resort cycle: Background to the legalisation of gambling. *Annals of Tourism Research, 5*(2), 238–251.

Storper, M. (1997). Territories, flows, and hierarchies in the global economy. In K. R. Cox (Ed.), *Spaces of globalisation: Reasserting the power of the local* (pp. 19–44). New York and London: The Guilford Press.

Strapp, J. D. (1988). The resort cycle and second homes. *Annals of Tourism Research, 15*(4), 504–516.

Timothy, D., Saarinen, J., & Viken, A. (2016). Tourism issues and international borders in the Nordic Region. *Scandinavian Journal of Hospitality and Tourism, 16*(1 Supplement), 1–13, doi:10.1080/15022250.2016.1244504

Torres, R., & Momsen, J. (2005). Gringolandia: The construction of a new tourist space in Mexico. *Annals of the Association of American Geographers, 95*, 314–335. doi:10.1111/j.1467-8306.2005.00462.x

United Nations (2015). Transforming our world: The 2030 agenda for sustainable development. Resolution adopted by the General Assembly on 25 September 2015. New York: United Nations.

Utting, P. (2015). Forewords. In Revisiting sustainable development. In P. Utting (Ed.), *Revisiting sustainable development* (pp. 1–15). Geneva: UNRISD.

Weaver, A. (2005). Spaces of containment and revenue capture: 'Super-sized' cruise ships as mobile tourism enclaves. *Tourism Geographies, 7*, 165–184.

Wolf, M. (2019). Why rigged capitalism is damaging liberal democracy. Financial Times, 18 September. Retrieved from https://www.ft.com/content/5a8ab27e-d470-11e9-8367-807ebd53ab77

Wong, C. S., & Kwong, W. Y. (2004). Outbound tourists' criteria for choosing all-inclusive package tours. *Tourism Management, 25*, 581–592.

Yle (2020). Business lobby calls for government action to save tourism industry. 3 September 2020. Retrieved from https://yle.fi/uutiset/osasto/news/business_lobby_calls_for_government_action_to_save_tourism_industry/11525296

Index

Note: **Bold** page numbers refer to tables and *italic* page numbers refer to figures.